Praise for *Mir(*

"*Miraculous Abundance* offers one of   nts for
earth-healing abundance I've ever   ie-blue
practitioners, hands in the soil, can offer the kind of ......,   abining
the best of all the earth-healing traditions and technologies — discovered on this
permaculture microfarm. A fantastic book with iconic potential. I couldn't put
it down."   —JOEL SALATIN, owner, Polyface Farm; author of *Fields of Farmers*

"*Miraculous Abundance* is a true marvel! Like Perrine and Charles Hervé-Gruyer's
amazing farm, their book blends science and anthropology, but it also mixes mem-
oir and travelogue to create a beautiful whole that will inspire the next generation
of farmers."   —NOVELLA CARPENTER, author of *Farm City*,
coauthor of *The Essential Urban Farmer*

"*Miraculous Abundance* is a dynamic combination of permaculture, biointensive,
four season, natural farming, and Amazonian farming approaches with exciting
practical goals to pattern after. The book is about healing ourselves and the Earth
in a post-carbon era. Worth reading for inner growth and outer growing of food,
compost materials, income, and soil!"   —JOHN JEAVONS, author; biologically
intensive farming specialist

"*Miraculous Abundance* tells the story of a pioneering permaculture market garden in
France. Small, highly diverse, highly productive microfarms are a critical part of
climate-change mitigation; their 'agroecological intensification' means we can grow
more on less land and reduce deforestation at the same time. Perrine and Charles
Hervé-Gruyer's book covers more than the logistics of their operation — it delves
into their philosophy and historical roots in French market garden history. A powerful case study of an intensive, commercial permaculture production system."
—ERIC TOENSMEIER, author of *The Carbon Farming Solution* and *Paradise Lot*

"In this lovely, hopeful book, an unlikely couple creates an astonishingly productive
edible landscape in Normandy, weaving together the insights, materials, and techniques of dozens of acknowledged predecessors. *Miraculous Abundance* is a modestly
written song of defiance, a demonstration that the world can readily feed its pro-
jected 9 billion with an agriculture that restores the biosphere."
—JOAN GUSSOW, author of *Growing, Older* and *This Organic Life*

"'Dare to imagine the new,' Perrine and Charles Hervé-Gruyer tell us. 'Take the best
of the many traditions of humanity, and the best of modernity, to shape a world
that has never existed.' These authors synthesize the best from multiple indigenous
cultures with successful patterns of modern small-scale farming to create a soaring
example and vision of a future — one in which human beings are an essential and
positive force helping to preserve the biosphere, and even a quarter acre can be a
full-fledged and productive farm yielding amazing agricultural bounty."
—CAROL DEPPE, author of *The Tao of Vegetable Gardening* and *The Resilient Gardener*

"*Miraculous Abundance* is absolutely the right book for right now. I don't know when I have been more encouraged about the future. . . . They are combining biointensive farming and permaculture to make a viable, diversified microfarm on test plots that are little more than two acres with the possibility of reducing that size down to as small as one fourth acre. . . . They use hardly any fossil-fuel energy at all, calling what they do the 'agriculture of the sun.' Along the way they provide solid evidence from sources all over the world to back up the conclusions they are drawing from their work, including achieving more healthful food, food security for the coming population increases, more jobs, effective sequestration of CO_2, and indeed a whole new world order that would insure better social stability out of the chaos we presently face." —GENE LOGSDON, author of *A Sanctuary of Trees* and *The Contrary Farmer*

"Can farming a tiny quarter-acre piece of land be sustainable, economic, and fulfill-ing? In *Miraculous Abundance*, Perrine and Charles Hervé-Gruyer tackle that very questions and answer it positively in the affirmative. This fascinating book describes the evolution of their farm from its beginnings in 2004, when the authors knew little, over the next ten years as they discovered biointensive agriculture, permacul-ture, forest gardens, and more. The authors are passionate about small, human-scale farming and the role it can play in the future, and they envisage a future with numerous small farms, enabling many more people to live on the land and lessen-ing the effects of climate change. Their farm in France now attracts farmers, chefs, and scientists and also hosts a school to teach how a diverse edible landscape can be created to both earn a living and make a beautiful space and a fulfilling life." —MARTIN CRAWFORD, author of *Trees for Gardens, Orchards and Permaculture*

"This book will be a source of inspiration and guidance for those striving toward an agriculture that is not merely sustainable but also regenerative and rewarding. Charles and Perrine are trailblazers, courageous visionaries who have drawn inspi-ration from sources as varied as 19th century Parisian market gardeners and Amazonian tribes people. As their method is a fusion, so too is the book; practical, historical, and philosophical in tone, it shows us how practicing agriculture as part of the ecosystem is not only economically viable but also spiritually fulfilling. We need people like the Hervé-Gruyers to show us what is possible in reality rather than just theory, and in sharing their journey, they tell an important story for the future of humankind." —CAROLINE AITKEN, permaculture teacher and consultant at Patrick Whitefield Associates; coauthor of *Food from Your Forest Garden*

"At their farm in Normandy, France, Perrine and Charles Hervé-Gruyer have cre-ated an inspiring example of how it is possible to intensively farm a small plot of land and produce an abundance of food while at the same time enriching the fertil-ity of the soil and the health of the people, plants, and animals that live there." —LARRY KORN, author of *One-Straw Revolutionary*; editor of Masanobu Fukuoka's *The One-Straw Revolution* and *Sowing Seeds in the Desert*

Miraculous
ABUNDANCE

Miraculous ABUNDANCE

One Quarter Acre, Two French Farmers, and Enough Food to Feed the World

PERRINE and CHARLES
HERVÉ-GRUYER

Translated by John F. Reynolds
Foreword by Eliot Coleman

Chelsea Green Publishing
White River Junction, Vermont

Editor: Joni Praded
Project Manager: Angela Boyle
Copy Editor: Laura Jorstad
Proofreader: Helen Walden
Indexer: Lee Lawton
Designer: Melissa Jacobson

Printed in the United States of America.
First printing March, 2016.
10 9 8 7 6 5 4 3 2 1   16 17 18 19 20

Our Commitment to Green Publishing
Chelsea Green sees publishing as a tool for cultural change and ecological stewardship. We strive to align our book manufacturing practices with our editorial mission and to reduce the impact of our business enterprise in the environment. We print our books and catalogs on chlorine-free recycled paper, using vegetable-based inks whenever possible. This book may cost slightly more because it was printed on paper that contains recycled fiber, and we hope you'll agree that it's worth it. Chelsea Green is a member of the Green Press Initiative (www.greenpressinitiative.org), a nonprofit coalition of publishers, manufacturers, and authors working to protect the world's endangered forests and conserve natural resources. *Miraculous Abundance* was printed on paper supplied by Thomson-Shore that contains 100% postconsumer recycled fiber.

Library of Congress Cataloging-in-Publication Data
Names: Hervé-Gruyer, Perrine, author. | Hervé-Gruyer, Charles, author.
Title: Miraculous abundance : one quarter acre, two French farmers, and enough food to feed the
     world / Perrine and Charles Hervé-Gruyer ; translated by John F. Reynolds.
Other titles: One quarter acre, two French farmers, and enough food to feed the world
Description: White River Junction, Vermont : Chelsea Green Publishing, [2016] | Includes
     bibliographical references and index.
Identifiers: LCCN 2015045172| ISBN 9781603586429 (pbk.) | ISBN 9781603586436 (ebook)
Subjects: LCSH: Permaculture. | Organic farming.
Classification: LCC S494.5.P47 H47 2016 | DDC 631.5/8--dc23
LC record available at http://lccn.loc.gov/2015045172

Chelsea Green Publishing
85 North Main Street, Suite 120
White River Junction, VT 05001
(802) 295-6300
www.chelseagreen.com

MIX
Paper from
responsible sources
FSC® C013483

For our children, Lila, Rose, Shanti, and Fénoua,
and for all the children of the world

For Eliot Coleman, Philippe Desbrosses,
François Léger, and François Lemarchand

You think you can stamp on that caterpillar?
All right, you've done it. It wasn't difficult.
And now, make the caterpillar again . . .

— Lanza del Vasto[1]

CONTENTS

FOREWORD

*M*y first meeting with Charles and Perrine Hervé-Gruyer a few years ago seemed more like a reunion with old friends than an initial encounter. Our shared passion for feeding the world with elegantly simple systems based on a profound intellectual reverence for the soil was evident from the first moment.

Their farm brought to mind a quote I had read from Amory Lovins, the energy guru. The writer of the article had trouble understanding Lovins's confident answers to her questions. She seemed to expect long and complicated replies with him bemoaning the great difficulty and potential insolubility of the problems. When she expressed this concern, his simple reply was "I don't do problems; I do solutions." Charles and Perrine's farm demonstrates that reply on every square inch. They do solutions. As a result, La Ferme du Bec Hellouin is the finest example I have seen of truly commercial permaculture/market gardening and goes far beyond the norm with the inclusion of fruit and nut trees. It is like a United Nations of all the best sustainable farming ideas.

Farmers like Charles and Perrine are among the beacons of light, autonomous independent examples of human beings who are not cogs in the industrial machine and are thereby able to experience the self-fulfillment of defining their own goal and working towards it. Their work allows the rest of the world to see that there is another life, there is another way. Located in France's Normandy region, La Ferme du Bec Hellouin has achieved widespread recognition. For their efforts to regenerate soil and land — which absorbs carbon — and to farm without the aid of fossil fuels, Charles and Perrine have been honored as Climate Heroes by the European group of that name. They were also recently featured in the French documentary, *Demain*, which had its debut at the COP21 conference in Paris. *Demain* celebrates the individuals who are pursuing solutions for the future.

La Ferme du Bec Hellouin is in the tradition of the farms supported by the Slow Food movement. Slow Food works to prevent the disappearance of traditional food products and the small farms that produce them. It is a noble effort. However, sometimes even those who should understand what is involved seem to miss the point. Slow Food is often criticized as elitist. But that criticism is totally blind to the real issue. It is not important whether you or I or anyone else in the United States ever gets to eat some specific artisanal food. What's important is that it exists, that there is one small corner of the planet still unconquered by Kraft or Nabisco or Monsanto, one little rural holdout inhabited by a few hardworking people who still know what quality is and have a passion for producing it. But, thankfully, La Bec Hellouin is not alone. All small farmers are that last unconquered hamlet. We are the ones who understand that from our farms will come the many new cheeses or tastier vegetables or unique grain products of the new world food order. These are the products of our soils.

When soil is used to produce at its full potential, the soil on our own farms can provide everything we need through efficient systems that take advantage of the synergy inherent in all the diverse pieces of the biology of the natural world. A fertile soil has the power to make the small farm ever more independent of purchased inputs and, thus, ever more independent of the corporate/industrial world. But the obvious question is this: if these systems work so well now and were so clear to our predecessors, why have they been ignored? Why have the benefits of increased soil organic matter and composting and crop rotation and mixed farming not been taught to farmers? That's probably because those superior production techniques can't be cheaply copied by industrial methods. Thus they are anathema to a bigger-is-better world. But we know differently. In the eyes of those of us who understand the benefits of producing food at the family-farm scale, this is the way farms should be run and the way life should be.

Fortunately for the planet, the movement toward real food, toward local food, toward food produced *with* care by farmers *who* care is the wave of the future. And it can't be stopped as long as we understand our advantages in working with the natural forces of the earth. Rediscovering the immutable value of the small farm is the first step toward a new agriculture for the 21st century and, possibly, a new world of the 21st century. The social and cultural influence of the productive family farm, celebrated for centuries by agrarian philosophers, can once again extend from the fertile soil under the farmer's feet far out beyond the boundaries of the farm itself.

Foreword

When I was first starting out many years ago, my skills and my agricultural philosophy were enormously influenced by the many competent French vegetable growers I was fortunate to visit. They were still connected to the age-old rhythms of the earth and were eager to share their knowledge. *Miraculous Abundance* continues that long tradition and will teach you as I was taught. Is this important work? I can't think of anything more so.

— ELIOT COLEMAN

INTRODUCTION

$\mathcal{I}$t all began when Perrine and I decided we wanted to spend our days feeling the sun and the rain on our skin, swimming in a river, and feeding our family with safe and vibrant food, cultivated with love by our own hands. So we became farmers, giving up city life for the Normandy countryside, embracing the land as something essential. Our dream was to live as close as possible to plants and animals. This path has been anything but easy, with plenty of disappointments, mistakes, and discouragements, but also loaded with dazzling happiness.

Today the view from the table where I write is an explosion of flowers and vegetables. The luxurious family garden stretches down a bank to the Bec river, which passes through our farm, La Ferme du Bec Hellouin. On the opposite bank lies our market garden. It looks nothing like a classic agricultural operation. Our inspiration comes from elsewhere: indigenous people from around the world and farmers from another era, but also the latest advances in natural agriculture. Our farm follows the practice of permaculture, an approach virtually unknown in France. Permaculture can be described as a box of smart tools that allows the creation of a lifestyle that respects the earth and its inhabitants — a practical method inspired by nature. And using nature as a model is exactly what we set out to do when we began this adventure.

We have painstakingly transformed a mediocre field into an edible landscape. The farm is a mosaic of intermingled ecosystems — ponds, islands, orchards, forest gardens, mounded crops, pastures. Fruit trees are everywhere. Animals and wild plants all seem to feel right at home. Generous abundance surrounds us. In the world of conventional agriculture, increasing productivity is normally at odds with respecting the environment. The two goals are viewed as opposites — as if nature were unproductive. But the verdict is in: Yields from our gardens surprise agronomists, and naturalists are shocked by the number of wild animals living in these gardens.

Nothing prepared us to become farmers. Perrine was an international lawyer; I was a sailor. We embarked on the project as complete beginners by experimenting, without any formal agricultural training, driven by a dream to work with our hands. We wanted to produce an abundance of delicious fruits and vegetables on a small piece of land using natural techniques. The disappointing results of our first few years led us to look for ways to adapt. We had no idea that our new profession would lead us into a fascinating worldwide investigation.

Our search for a better way introduced us to some of the original organic farmers of France, but also to Americans, who were bursting with creativity and helped us discover the rich Parisian market garden tradition of the nineteenth century. It became an amazing journey through space and time — in the garden by daylight and with books or the Internet by night, a rake in one hand, a computer in the other. It was an investigation that took Perrine from Japan to the United States, Cuba, and England in search of innovative solutions. As a result, the farm has become the setting for multiple experiments driven by the same goal: producing more in less space, with complete respect for nature.

We have not invented anything. Like bees, we have foraged from a wide variety of sources, off the beaten path, taking — more and more resolutely — the opposite approach to industrialized agriculture, which artificializes nature. We have taken advantage of services generously provided by ecosystems. After all, doesn't nature offer plants everything they need to grow: energy from the sun and rainwater, nitrogen and carbon from the atmosphere, mineral salts from the mother rock, and the extraordinary work of organisms living in humus?

Gradually, we stopped believing that we are the ones who grow the plants. The potential of a plant is contained in the seed; the mission of the soil is to ensure its germination and growth. We are the modest assistants of these life forces. Our mission is to provide plants the most favorable conditions for their development. We are the servants of the earthworms!

Little by little, we have come to define the Bec Hellouin farming method as a permaculture approach based on a risky plan to put human labor at the heart of agricultural production. Yes, we have devised a system rich in labor: a snub to the agricultural industry, which continues to replace people with machines and fossil fuels. Is this a crazy gamble at a time when labor is expensive and agricultural products are generally selling for cheap? Certainly, but permaculture presents a solution to the problem. The human hand becomes an asset when it takes on a task that a machine cannot easily achieve — cultivating intensely alive spaces, lovingly caring for the soil and plants, mingling crops, and packing them densely together to fend off weeds. We came to realize that our soil had become so fertile and our production so consequential that agronomists from France, Japan, the United States, Brazil, even Africa, began to visit the farm.

From all of these developments, a conviction was born: Microagriculture can be an innovative solution to many environmental and social problems. It is an alternative that will become, over the years, more and more precious because our food is heavily dependent on oil. It currently takes an average of 10 to 12 calories of fossil fuel to produce 1 calorie of food on our plates,[1] which is so outrageous that the English author Albert Bartlett wrote, "Modern agriculture is the use of land to convert petroleum into food." Oil, inexorably, we will have less and less of, and the price will always rise . . . yet we all intend to continue eating![2]

Bountiful production on a small plot, while creating jobs, enriching the environment, increasing soil fertility, storing carbon, and conserving biodiversity? That might seem too good to be true, and is the exact opposite of what humanity is currently achieving. People today continually drift farther and farther away from nature, replacing life with technology in pursuit of a consumerist dream that is made possible by the senseless waste, in just three or four generations, of energy that took millions of years to form.

It is precisely because permaculture uses nature as a model that it opens up exciting prospects. Permaculture is the exact opposite of the current logic: It represents a new paradigm, a new "software" that seeks to reconcile human beings with the earth.

In these unprecedented times of ecological and social crisis, as we enter a period of declining energy that will shake the very foundations of our civilization, permaculture helps us imagine a future rich with an abundance of essential goods — simply because it is inspired by nature, which has always been able to generate overflowing ecosystem vitality, even in resource-poor settings.[3]

Food accounts for about one-third of greenhouse gas emissions.[4] Can we hope to feed the world while restoring the planet? The possibility of allying permaculture with small family and subsistence farms (which are still the dominant model; 90 percent of farms in the world are less than 2 hectares/5 acres) persuades us to answer with a resolute *yes!* This approach, even though it is still in its infancy, can be a source of inspiration for conceiving the alternatives of tomorrow.[5]

An agronomic research program is under way at La Ferme du Bec Hellouin. We designed it with François Léger, head of France's SAD-APT (Science for Action and Sustainable Development: Activities, Products, Territories) research unit, a team of about sixty researchers at the National Institute for Agricultural Research, also known as INRA, and AgroParis-Tech. François has always been interested in experimental farmers, convinced that they are the lifeblood of innovation. This study, called "Organic Permaculture Market Gardening and Economic Performance,"[6] attempts to answer the following question: Can 1,000 square meters of diverse vegetable crops, grown with the Bec Hellouin method, provide a full-time occupation for a farmer?[7] We have isolated an equivalent plot in our gardens and are recording everything that goes into it and everything that comes out of it, right down to the last bunch of radishes. Since the first year, we have been able to demonstrate that this goal is viable.[8]

Although we are still early in the research program, we were able to verify what John Jeavons suggested in the late 1970s: It is possible for an experienced market gardener, working by hand, to produce, in an equal amount of work time, as many vegetables as a farmer equipped with a tractor. The father of biointensive agriculture, Jeavons is one of our major sources of inspiration from the United States. This type of bio-inspired agriculture — that is, agriculture inspired by life — will, we hope, contribute to making it possible for a new generation of farms to feed the human population while restoring the environment.

We came to realize that, when we touch the earth, it connects us to everything that makes up human life: food, of course, but also health, landscapes, employment, the economy, the art of living together, and even what we hold most dear — our emotions, our presence in the world, our relationship with life. And we discovered that our business of farming on a small piece of land in a Normandy valley can have an impact on all major contemporary issues, including food security, protection of biodiversity, hunger, and global warming. This prospect fills us with hope and a desire to forge ahead. If we want to live sustainably on this planet, a growing number of people will have to reconnect to the land and produce food for themselves or the community. As bio-agriculture pioneer Philippe Desbrosses writes, "We will re-become farmers."[9] A society cannot survive with only 2 to 3 percent of the population farming. But the farmers of tomorrow will not come from the agricultural class that has been reduced to near extinction; they will come from cities, offices, shops, factories, and more. One thing is certain, they will not return to the earth using conventional models from the recent past. We need to invent new ways of being farmers in the twenty-first century. The farmers of tomorrow will be the guardians of life. Their farms will be places of healing, of beauty, and of harmony.

Our farming journey is still just at the beginning. We will have much more to do, and still have many more questions than answers. However, hundreds of people come each year to learn at La Ferme du Bec Hellouin and ask us insistently to tell the story of this adventure. Now, with the first data from the study mentioned above, we believe that the time has come to share this approach more widely, although most of the road still lies ahead of us.

Our journey is the thread for this book — a human adventure makes it more alive, more personal — but our main goal is to share our passion for permaculture, its implementation at La Ferme du Bec Hellouin, and the innovative experiences that have inspired us. We will journey to the four corners of the world to meet pioneer farmers.

Explaining how our form of bio-inspired agriculture has progressively developed leads us to also discuss some of our experiences prior to the creation of the farm, which includes my years spent learning from the indigenous people of the Amazon, Africa, and elsewhere. We have not focused much attention on the dominant paradigms of the West because our inspirations come from other cultures — cultures that emphasize respect for life and preservation of the environment. Throughout the story we will be visiting those whom I call my barefoot mentors.

In the last part of the book, we envision a future in which agriculture is based on innovative concepts: the forest garden, the microfarm, ecosystems of the microfarm, and the cooperative agrarian system, to name a few. These scenarios may open the realm of possibility for all those who dream, in one way or another, of becoming farmers — but also for elected officials and policy makers who wish to develop, even in the heart of cities, quality organic agriculture.

When the scientific study happening at La Ferme du Bec Hellouin is complete, we will write a practical guide sharing more how-to details on what we have learned about gardening and microagriculture permaculture.

May these pages awaken creativity and the desire to put hands in the dirt!

ONE

Pupoli's Canoe

Industrial market societies have material resources, physical health, social organization, scientific knowledge and techniques, which, collectively, allow them to dominate the world. But where is the daily happiness? Their knowledge of destiny? Their connection to the dead? Nowhere. Their souls are searching in vain for a den.

— *Jean Ziegler*[1]

Every civilization is an alliance with the universe. The universe is never a static ensemble; it's what man makes of this alliance that matters.

— *Robert Jaulin*[2]

*A*ntecume Pata is a small village of the Wayana people, located on an island in the Litany River, on the border between French Guiana and Suriname. The river is wide there, churning with rapids. The tumultuous waves foam up as they crash against the black rocks. On the banks, the Amazon forest stretches as far as you can see; the only clearing is that opened up by the Indians to build their huts.

I have returned to Antecume Pata, an important place in my life, many times, watching the Indians that I met as children grow into adulthood. On each trip my trust and friendship grew with the Wayanas, a modest and discreet people at first, but so endearing and humorous as intimate friends.

Monkey Brother

Pupoli was one of my companions. His father, Yoïwet, and I were very close. Yoïwet even gave me a nickname that he also applied to himself: *yepe baboune* (monkey brother). The exchange of nicknames is for Wayanas a great sign of friendship. It took ten years of travel to the end of Guyana for such a bond to be born.

One day I had an adventure, seemingly innocuous, with Pupoli that left an indelible mark on me. Though it was in 1995, I remember like it happened just yesterday. Pupoli, then a slight boy about ten years old, invited me to go fishing in his canoe. We set out, each dressed in a kalimbe, a single band of bright red cloth placed between the legs. Carved from a single tree trunk, Pupoli's canoe was no bigger than a toy, perfectly unstable, the hull flush with the water. I felt like all I had to do was shift my elbow a bit and we would capsize. Fortunately Pupoli was more comfortable than I was, playing energetically with the paddle, his little fishing rod at the bottom of the canoe with its line and some worms for bait.

The young Wayana live the freest childhood imaginable; they learn indigenous life by using tools identical to those of adults, except they are made to their size. Their ease in nature is amazing.

We headed up the Litany River, penetrating the Amazon forest, a sumptuous garden of Eden. Soon we heard a powerful roar: We approached an impressive waterfall that blocked the Litany across its entire width. Despite the strong current, the boy navigated the flow of the river without any apparent effort. I wondered how far the brave and adventurous Pupoli would take us. The child stopped only a few meters from the falls. There he laid his paddle at the bottom of the canoe, unrolled his line, and began to fish. It all seemed so simple — a child's game. But by what miracle was the small Wayana boy able to overcome the strong current so effortlessly? I watched, fascinated. Pupoli maneuvered in the river simply by keeping his canoe in the countercurrents generated by rapids, skillfully gliding from one rock to another. The fragile little boat would turn precisely in a narrow calm zone, perhaps the only spot within the entire width of the river where there was a vein of countercurrent. If we had been just a few meters off, we would have been swept away by the rushing water against which resistance would have been in vain.

And a bite! Just like that, Pupoli pulled in a piranha with red eyes and fierce teeth. With a swipe of the machete, he split the skull before throwing it to the bottom of the canoe, for fear that the fish would use its teeth on our toes.

I was overwhelmed with admiration by the ease and obvious skill with which this child had played the seemingly indomitable river. Achieving such elegance required an intimate knowledge of his environment. While fishing, I reflected on the lesson Pupoli had unknowingly just given me. A current always causes a countercurrent. And as the current becomes more powerful, the countercurrent likewise becomes stronger.

If Pupoli managed to position himself in a favorable vein, he accomplished his goal, even though the balance of power between the river and his little arms was totally unfavorable.

I felt an immense joy rising up in me. I had until then seen our world as the great river: terribly powerful. And I often found myself swept up against my will by the current, unable to resist. Modernity takes us along without asking our opinion, and no one really knows where we're going. Yet this world that is so incredibly powerful generates countercurrents: If I learned to identify them, I realized, I would not have to struggle, to exhaust myself in a losing battle. By positioning myself in my proper place, I would be able to carve out a path aligned with my heart and my dreams.

Permaculture: Inspired by Nature

The defining trait of the modern Western world is hypertechnicity, a pursuit of material progress. Despite undeniable advances in countless areas, this form of progress, as it has been developed to date, provokes a rapid and massive destruction of the biosphere. We continue to artificialize nature, replacing life with technology.

It is the dominant current — strong, rapid, and terrifying.

But the countercurrents exist, everywhere: lively little veins of water that bring us hope. All around the world, millions of people of goodwill are striving with all their force to invent lifestyles that are respectful of people and the planet.

Permaculture is one of those countercurrents. This approach recognizes the essence of life. It aims to follow the school of nature, and be fertilized by her. For the past 3.8 billion years, life has colonized planet earth, creating favorable conditions for the emergence of increasingly more complex forms of life. Obviously, this process took place without human intervention.

Permaculture is a bio-inspired approach, and in this sense it is exactly the opposite of the contemporary dominant current that weakens the biosphere. It represents a new paradigm for those looking to heal the earth. Its objective is to design human infrastructure that functions, as much as possible, like natural ecosystems. Permaculture allows everyone to invent a lifestyle that suits them, in harmony with the planet. The method was formulated in the 1970s by Bill Mollison and David Holmgren, who were heavily inspired by observing Aboriginal peoples. "To injure a tree was to injure a brother; this outlook, then, is one of sophisticated conservation. Can you cut down a brother and live?" asked Mollison.[3]

Permaculture is based on an ethic that's simple in its formulation, but demanding in its implementation:

> Take care of the earth.
> Take care of the people.
> Equitably share resources.

This book is not intended to describe permaculture in its entirety. To understand permaculture systematically, you can refer to the books listed in the Resources section.

Permaculture and Biological Agriculture

These pages relate to permaculture through our experience as small farmers. Practicing our profession, we found that the concepts of permaculture are still little known and rarely applied in the world of biological agriculture. New permaculture farms are rare in many parts of the world, which is a paradox because, from its origins, permaculture has been concerned primarily with food production. This focus on food production sometimes causes misunderstanding: Permaculture is reduced to a natural gardening super-method. Yet permaculture is not a collection of agricultural techniques. Its potential goes well beyond that, to a conceptual system that is likely to fertilize all our human achievements. Some confusion may arise in the mind of the reader unfamiliar with the world of agriculture. We are often asked about the difference between permaculture, organic agriculture, and agroecology. In a few words:

- **Organic agriculture** is a branch of agriculture that prohibits the use of synthetic particles (fertilizers, herbicides, and chemical pesticides) and promotes high standards of respect for plants, animals, and agricultural systems. It is regulated by a set of official rules and subjected to checks and approval.

- **Agroecology** integrates environmental and social considerations in order to meet the food needs of human communities while respecting farmers and nature. Its definition is less clear than that of organic farming, and it is not subject to specific legislation. Agroecology does not necessarily exclude the use of synthetic products.

- **Permaculture** is based on extremely close observation of the functioning of natural ecosystems. Farmers who aspire to an agriculture that is as natural as possible will find great satisfaction in applying the concepts of permaculture to their own operation; according to our experience, permaculture takes all of the great attributes of past practices and allows us to improve further upon them.

So the objectives of permaculture are wider than those of organic agriculture and agroecology since it transcends the agricultural sphere. To design a truly ecological installation (a farm, company, or city, for instance), permaculture integrates all the good practices of organic agriculture and agroecology, and crosses them with "green approaches" from other disciplines (like renewable energy and green building). There are no conflicts among organic agriculture, agroecology, and permaculture. On the contrary, simply a difference in nature.

Developing bio-inspired agriculture can help to sustainably feed the world. The food challenge is significant: Today 842 million human beings are suffering from hunger.[4] That's one in eight people. Every eleven years, the world population increases by a billion more people. According to a UN report, world food production should increase from current levels by 70 to 100 percent by 2050 if it intends to meet the food needs of a growing population.[5] However, since the 1960s, a third of arable land on the planet has disappeared because of erosion, exacerbated by the rise of industrial agriculture and artificial contaminants in the soil.[6] An area the size of Italy is lost annually.[7] The food challenge therefore goes well beyond the borders of the world of farmers and concerns us all. Permaculture has much to contribute to the ongoing discussions about the creation of productive, autonomous, and resilient agricultural systems. Designed as ecosystems, small-sized gardens can have a surprising amount of productivity. The Dervaes family, in California, generates an annual revenue of $20,000 (about 14,500 euros) with a garden of 360 square meters, providing income to the father, his son, and his two daughters while nourishing the local community.[8] Examples of similar achievements are numerous.

At La Ferme du Bec Hellouin, we get our energy primarily from the sun, with the lightest possible use of fossil fuels. We experiment with different "best practices," some going back to ancient civilizations; others, on the contrary, at the forefront of innovation. We reach production levels that seem barely believable to specialists with simple and natural practices that are good for humans and the environment. According to naturalists, biodiversity is increasing in our gardens; a true regeneration of the biotope is under way. We have come to understand that one can be a small farmer and actively participate in the healing of the biosphere.

Interior Landscapes, Exterior Landscapes

Our farm is exploring alternative tracks, setting our roots in the desire to influence the main current of life, a desire encouraged by experiencing communities that have made choices very different from our modern Western world's. Meeting indigenous people around the world has fertilized our imagination. Their cultures have undergone as many thousands of years of evolution as ours. They have simply made different choices. We prioritize *having*; they are in search of harmony. We live in the short term, they are devoted to the long term. We see ourselves as separate from nature, they see themselves as part of the wider community of the living. These first peoples have much to say to today's people, as they can do what we have forgotten: live in harmony with nature. They remind us that the way we shape the landscape we live in reflects our inner landscapes. What is our conception of happiness? A Native American once said, "The white man, when he dies, wishes to leave money to his children. The Indian, he wants to give them trees."

To be inspired by their wisdom, however, does not mean that we have to abandon our culture, and most notably the considerable scientific advances of the past decades. Our knowledge of biology doubles every five years![9] At

La Ferme du Bec Hellouin we try, in our approach to agriculture, to combine science and conscience, intuition and rigor, convinced that agriculture can become an art — a body of scientific and technical knowledge fertilized by intuition and human creativity.

In other words, it amounts to taking the best of both worlds, the best of tradition and the best of modernity. We are discovering, while creating our farm, just how exhilarating it can be to achieve a synthesis of original solutions from different cultures and eras.

TWO

Around the World

To live on the earth as a poet or an assassin?

— Paul Virilio[1]

When we burn grass for grasshoppers, we do not ruin things. We shake down acorns and pine nuts . . . The white man, he turns over the ground, kills the trees, destroys everything . . . He explodes the rocks and leaves them scattered on the ground . . . How can the Spirit of the Earth love the white man? Everywhere he touches, it leaves a wound.

— a Wintu woman[2]

*I*t took thirty years for me to become a farmer. When I was a teenager, part-time work with a family of Normandy farmers foreshadowed my vocation: I wanted to cultivate the earth, live outside, free, under the rain and the sun. But people told me that, as a Parisian, I would struggle to become a farmer, and my thirst for discovery took over. Working as a sailor was a nice alternative. At age twenty-one I took to the sea to try to understand this vast world and find my place in it. I had a sailboat, a school vessel named *Fleur de Lampaul*. Crews of teenagers and scientists joined me on each excursion, and we traveled all the oceans of the world.[3]

At first, we studied marine mammals, coming across them in their element and relishing our improbable encounters. Sometimes, when diving, a rorqual or a sperm whale the size of a bus would approach the little swimmers that we were, and then turn on its side to contemplate us with his impassive eye. In these magical moments, we seemed to cross an invisible border, recovering some of the friendship between humans and animals that should reign on this earthly paradise. This symbiosis was exactly what I wanted. But we cannot spend our life underwater. After a few years, I realized that humans were no less interesting than marine mammals.

The *Fleur de Lampaul* then embarked on a fifteen-year journey to meet the first peoples: Native Americans, African tribes, Australian Aborigines, Papuans of the Vanuatu. On each journey, we shared their lives as intimately as possible, placing ourselves humbly in their school. In their company — dressed in a djellaba in the Sahara, a kalimbe in the Amazon, a sarong in the Marquesas Islands, or even a penis sheath (an ecological garment par excellence, its only drawback being that it is a bit itchy) — we discovered how they live in environments as diverse as coral or volcanic islands, primary forests, deserts, or mangroves. What difficulties did they face? What solutions had they developed?

My Masters of Nature

We received from them unforgettable lessons of humanity and wisdom. I had great barefoot master teachers: King Sylva of the Tabanca Bane Ijun, in West Africa's Bijagos archipelago; Mimi Siku, Wayana hunter from Guyana; the sahila, or chief, Pedro Hakin, of the Kuna people of the San Blas archipelago off Panama; to name a few. Their teachings are deeply etched in my heart. Living among them was, every day, an applied ecology lesson. I remember having walked in the Amazon rain forest in the company of Kalina children. When they gather an edible plant with the swing of a machete, quite naturally and without a thought, they replant cuttings the whole length of the trail. Anthropologists have found that the last Amazonian nomad tribes were not traveling at random, but followed itineraries well defined from generation to generation, from one watering hole to the next. By replanting nuts and cuttings of useful plants along their discreet pathways, they created edible corridors. The resources were available in abundance, life was made much easier, and they had plenty of free time to share together. Is there a gentler way of living in harmony with the land?[4]

I have a wonderful memory of our stop in Bali. We shared the daily life of a peasant family from the lowest caste, the Sudras. Gadeh, Komong, and

their children lived in a small house without water or electricity on plat-
forms in the middle of rice paddies, in the flow path of a volcano — a
landscape so beautiful that it takes your breath away. Gadeh and Komong
were farmworkers. They had to sell their own paddy to pay for the crema-
tion of their grandmother, so they worked during the day for the wealthier
farmers in the area, for a pittance. We slept on the floor and took care of our
needs in the irrigation canal. Yet what fulfillment in their way of life! The
children enjoyed music and dancing daily. At the end of the day, all families
took their bath in the river flowing in the hollow of the valley, the women
on one side, men on the other, naked, soaping up in the middle of the most
beautiful bathroom of the world. Then we dressed in clean clothes before
prayer, celebrated by Komong while the family gathered in front of a stone
statue. A frangipani flower over the ear here, raven hair covered with a
colorful turban there, faces that were serene and bright. The evening meal,
made with all the products offered by the surrounding nature, was a summit
of gastronomy and refinement.

Meeting the great pioneers of ecological thinking was also pivotal.
Reading their works during long crossings was enriched by spending time
with some who had agreed to lead some of our expeditions: René Dumont,
the agronomist rebel; Théodore Monod, the humanist who studied the
Sahara; Hubert Reeves, whose works have made us consider the cosmos
with new eyes. The lecture on the stars given by Hubert while the crew lay
out on deck under a starry vault was a great moment of science and poetry.

A Drop of Water on an Orange

When we are children, we believe that the world is immense. As adults,
many of us keep this belief, culturally reinforced by the difficulty of traveling, until recently. But since the 1960s, photographs of earth from space
have marked a turning point in the collective perception of our planet. So
this is it, the earth, this little blue garden lost in the stars? How fragile she
seems, looked down upon from space. As astrophysicist Hubert Reeves has
explained, the presence of life-forms on our planet is a pure miracle, or
rather an uninterrupted succession of miracles for billions of years. The
process of moving from the intial chaos to the incredible complex organization of our biosphere was extremely tenuous.[5]

Circumnavigating the globe in our heavy, slow wooden sailboat, I came to experience — physically — the tininess of the earth. Once, three years after leaving our home base off the coastal island of Yeu, we saw it appear again on the horizon, and I could hardly believe that the trip was already over. I understood on a gut level why our every action has so much impact, why the biosphere is collapsing so quickly, after decades of unbridled industrial "progress." "You can not pick a flower without disturbing a star," wrote the poet Francis Thompson.[6]

The biosphere is extraordinarily small. The thickness of the living part of our planet runs from just a few meters under our feet to just a few kilometers over our heads, where insects and seeds are carried by high-altitude winds. Certain bacteria, too, are equipped for long trips up there, where they can still reproduce and, incidentally, serve as condensation nuclei for raindrops, allowing them to come down. Yet we see the biosphere as much larger than it is. This is because from our birth to our death we are immersed in the heart of this thin film of life. The extreme rarity of that life on this planet goes unnoticed by us. Scientists tell us that the proportion of the biosphere to the planet can be likened to a single drop of water spread on the surface of an orange.

Over the years, I fell totally in love with this world we have the honor to live in, the only living planet known in the midst of the cosmic desert. Gardening this planet is an immense privilege and an equally great responsibility.

Asphyxiation of the Planet

During those years of traveling, we often visited some of the same places from time to time and saw rapid degradation of ecosystems. Desertification, destruction of primary forests and of mangroves, loss of coral reefs, rising sea levels, unbridled growth of cities: We witnessed the asphyxiation of the biosphere.

Passionate about diving, I loved filming the coral. But during our tour of the world, we searched in vain for intact coral reefs, even around remote islands of the Pacific and the Indian Ocean, the Marquesas, the Tuamotus, and the Maldives. Almost all the reefs are victims of ocean warming. I never saw the same luxuriance of coral reefs I had observed during my earlier sailing years. If my children one day make the same trip, there will be only the crumbs from the flamboyant past. Such realities caused the president of the Republic of the Maldives to make a plea to our band of travelers: He wanted us to unite all of our efforts to curb global warming. The islands of the atoll are already feeling the dramatic effects of rising sea levels. These paradise-like outcroppings, just 90 centimeters (3 feet) above sea level, are doomed to disappear, or at least to become uninhabitable. Our fishermen companions from the atoll of Fehendoo have already decided which country to migrate to. The human being has become, in barely a century, a force driving the evolution of the planet as powerfully as geological forces do. That's why some thinkers have dubbed the times we live in as the Anthropocene, or age of humans.[7]

These successive realizations did not have the effect of lifting morale. When looking at the state of the world, there is reason to be pessimistic. Twentieth-century wars have killed more people than all previous wars combined; more people died of hunger in the early twenty-first century than in the Middle Ages. I experienced these trips with my throat increasingly tight.

But one day, a phrase from the will of Abbot Pierre hit home. In the evening of his life, this great man advised us to stop lamenting: "Our old world is dying, certainly, but a new world is being born." These few words radically changed my perspective. I then made the choice to put all my energy in building the world to come. Since that moment I've felt much lighter and more joyful. That's why this book deliberately focuses on concrete proposals rather than denunciation.

In Search of Harmony

In two decades of navigation, the crews of the *Fleur de Lampaul* and I shared the lives of countless families and communities. The indigenous people we met opened the doors of their huts, their bungalows, and their tents, without hesitation. I wanted to live all of my life in such harmony. I thought about starting a family among the American Indians. But the worlds we discovered at the whim of these visits were not mine. I, too, had a culture, roots that I could not deny. Eventually, I sold my beloved boat. But once I did, I felt adrift: How could I stay true to what I had received from my barefoot teachers and find my way in a Western world that was so often materialistic and predatory? I missed the large open spaces, the softness of tribal communities.

Becoming a farmer was, for me, the answer. It was not an obvious move, but took shape gradually over the years. Much time passed before I allowed myself to live that teenage dream. At the time of this writing, despite all the difficulties of this path, I feel finally in profound agreement with myself, in my rightful place. Like Pupoli. Creating our farm has been an even more exciting adventure than sailing around the world. The farm, however, often seems like a boat: We are masters aboard our little universe, nose in the air to judge the direction of the clouds. Day after day we make our own choices and bear the consequences. The farmer, like the sailor, is a free man.

Often, after an intense workday, I dream of the tribes with whom a part of my heart remains. I now realize the extent to which my exchanges with native peoples have deconditioned me — deformatting my once Western-dominated thought and opening me up to other paradigms. My goal has been to cross what I've learned through our farm experiments, through lectures, and through exhanges with agronomists and naturalists who visited the farm with the teachings pulled from these years of navigating. I am steeped in these moments shared with tribal communities, of these immersions at the heart of unviolated ecosystems.

THREE

From Dream to Reality

Child, I knew giving. I lost this grace by becoming civilized. I was leading a natural life, whereas today I saw the artificial. The least pretty pebble was valuable to me. Every tree was an object of respect. Today, I admire the white man with a painted landscape whose value is expressed in dollars!

— Chiyesa[1]

A thatched-roof house, a river, 6,500 square meters of Normandy land, and a head full of dreams!

Our first spring on the farm was dotted with small miracles. In February 2004, we had spent a month moving into the cottage, alternating weeks in Normandy with weeks in Paris. One evening at dusk, we arrived from Meudon, just south of Paris, and Lila, seven, my oldest daughter, ran out to the meadow, then back in a hurry: "Papa, there is a little head! There are three sheep!" Sure enough, a little black lamb was just born to one of the Ouessant sheep we had bought the previous weekend. It was trembling on its large hooves, a white spot between its ears.

Then came March, when Rose, four, arrived shouting: "The chicks! There are chickens in the henhouse!" Surprised, we discovered a hen trying to gather under its wings a dozen unruly little balls of feathers. We had not seen the hatching; it was hidden in a dark corner of the barn.

By May the warm evening breeze was sprinkling the grass with millions of delicate white, rosy petals from the apple trees. The water of the Bec river, reflecting the first stars of the night, would carry them off to the sea.

And then came June. Up in a cherry tree, we filled baskets with child-like pleasure, nose in the branches, skin against bark. "Dreadfully sentimental, your story! Sounds like *Little House on the Prairie*," quipped Perrine one day. I agreed, yet these little things were all shots of oxygen, ephemeral pleasures offered day after day. The thatched cottage in Normandy was the place of our rebirth. The port after the storm.

Selling my boat and turning my back on the sea after twenty-two years of circumnavigating the world, I had turned the page on a big piece of my life. That was followed by a divorce, joint custody, and three years in Paris with the feeling of being a fish out of water. When the dream of farming took shape, I clung to it as if it were a rescue buoy. It drew us forward, my daughters and me. Le Bec Hellouin was a village that I had loved since childhood, just a few steps from an ancient abbey. We purchased and restored our once dilapidated cottage there. Soon came the sheep and chickens. Ponies, goats, rabbits, and a pig arrived to keep company with the sheep. Geese, ducks, guinea fowl, and turkeys rounded out the farmyard. The cottage was not a "real" farm at this point — just a neo rural fantasy (for proof, each animal had a name and we did not eat them). It looked like the houses in the French comic book series *Sylvain et Sylvette*, or the storybook *Caroline at the Farm*. Even so, it was where, eventually, we would let our roots grow and try to be happy together.

The most beautiful miracle of the year 2004, though, was Perrine. She was there when we slept for the first time under the thatched roof. And she has been there every day since, the farmer of my life.

From Tokyo Skyscrapers
to the Banks of the Bec River

Nothing, absolutely nothing, in her past hinted at Perrine becoming a farmer — except, perhaps, atavism; her family's ancestral roots are Italian and rural. Her grandparents had come to France to seek a better fortune in the city of Arras between the wars, and that's where she was born and raised. When she was very young, she was anything but passionate about nature: Touching the ground was disgusting, so she refused to sit on the grass at picnics.

But Perrine recalls a vacation at the family's mountain village in Italy. "I spent hours in the forest with my dog, climbing, making dams in the stream, and imagining adventures. These woods were my Amazon forest. They are why, like Idefix [the popular French cartoon dog who cries when trees are cut], I hate to see felled trees; I am almost unable even to prune them. Trees were my playmates."

The great business of her youth was sports. She joined the local basketball club at six years old and did not put down the ball until we moved to La Ferme du Bec Hellouin. She managed to reconcile schooling and brilliant law studies with evening training sessions, and games on the weekend. Sports have always been her way of immersing herself in her surroundings. Perrine is the type of person who engages fully in everything; she can move mountains and asks for nothing in return. Her friends call her Zorotte (Zorro, in female form) because she dreams of saving the world. I can attest that this dream is still alive and well.

With a diploma in both law and economic development, Perrine moved to Japan and fell under the spell of that country. She worked three and a half years in a law firm, while volunteering with the UN High Commission for Refugees, and also managed the legal department of a multinational company in China. This meeting with the Far East affected her profoundly. At thirty, having discovered meditation, the art of massage, and a way of looking at life much different from our Western one, Perrine came to wonder if there was meaning to working day and night to earn millions of dollars for a multinational corporation. She left her job and returned to France, looking for a new direction.

When we met, we were so different from each other. I was fifteen years her elder (this has not changed) with two children; she was free as air and eager to help change society. I had left my boat, and she had left skyscrapers. I wanted to put down roots; her dream was to travel again. Perrine was as comfortable as a fish in water with modernity; I wanted to live like the Indians.

The Road to Independence

Still, Perrine was well suited to life in the thatched-roof house at La Ferme du Bec Hellouin. It was as if she were naturally disposed to this new existence, surrounded by animals she loved. We had met while we were both studying to become psychotherapists. One year later, we married. For our honeymoon, we camped on a nearly deserted island. Perrine had never slept under the stars.

We both worked very hard during the three following years to develop our piece of land into a place of autonomy, able to feed ourselves and our children with healthy products. We planted fruit trees in large numbers, and created the first garden. In the first year the garden was full of vegetables and we were proud to offer full crates to our friends. "Not bad for a sailor!" shouted our gregarious neighbor, the elderly Father Autin, who, from his vantage point of eighty years, is the historian of the valley.

Eventually, a neighbor sold us a plot of 1.2 hectares (3 acres) on the other side of the Bec river. We built a bridge to connect the two shores, and it plays a big role in our farm; We cross it dozens of times a day, each time with an admiring look at the trout swimming in the clear water. The Bec is actually a canal dug in 1450 to bring water to the nearby abbey. Bec meant "river" in the Viking language, and the village's name, Hellouin, derives from the name of the knight Herluin, who founded the abbey in 1034. The Bec is lined with hewn stone and dotted, over 3 kilometers (1.9 miles), with seven water mills. The Bec valley today has two official distinctions: It is part of the national historic district surrounding the abbey, and also located in one of Europe's Natura 2000 protected areas. As time passed, we restored the riverbanks and created a beach where we bathe in the company of trout and geese. In grassland around the farm, we dug ponds that quickly populated with aquatic plants and frogs. We learned to make bread, and built an old-style oven out of stone and cob. (A very satisfying project: Build the oven, knead the dough, light a fire, bake the dough — so many gestures that taste of life's essentials. During cooking, the smell of warm bread spreads throughout the inner orchard.)

Father Autin sold us his cider press. "It is like new," he said. "It dates from 1948!" We restored it, and one fall when the orchard transformed into thousands of yellow and red pearls we learned to brew cider. The children washed apples; we ground them and filled the press. Then, with several turns of the wheel, juice began to flow, syrupy and amber-colored.

A few months later, in the cool of the cellar, we put up our first bottles of cider. Rose, sitting on the end of the barrel, filled the bottles. Lila sank the stopper with a tool nearly too big for her. Perrine and I laid the wire over the caps, and then there was the obligatory wait for the cider to "make foam" before tasting.

There were also jams, syrups, medicinal plant crops, teas, and other experiments. Not everything worked the first time around. There was misfiring. The cereal cultivation trials were particularly inconclusive. But as the months and years passed, we discovered the immense satisfaction that comes with eating almost entirely from our own production. When you have tasted the wonderful flavors and freshness of vegetables just harvested, you definitely do not miss the supermarkets.

The Draft Horse

All these experiments made us want to go farther. We bought Lou, a magnificant Normandy cob, one of nine French breeds of draft horses. I took a course on plowing with draft horses, and soon we were making hay bales in our new hay field. When the hay had dried, Lila and I transported cartful after cartful to the barn until nightfall. One of the joys of farming is that it can often be done with our children. They take great pride in being able to do useful work, in collaboration with adults.

Working with that first draft horse was a source of many joys, and sometimes stresses, too. After two years we chose a small Merens horse, Winick, who was more suitable to our needs. He has proven to be a faithful workmate.

By the time we had to build a shelter for the animals, the monks and nuns who surround the farm had sold us part of their land — 12 hectares (30 acres) of woods and a 2-hectare (5-acre) fallow over by the the abbey, about a kilometer (0.6 mile) away from the farm. We rode up the hill toward the abbey with the horse, cut down pine trees, and headed down with the cart full. Then we debarked the logs before building the shelter. Perrine's belly was swelling all the while. Until the eve of her confinement, she debarked her two trunks per day. I admired her indomitable courage. Perrine was investing in the farm with the same tremendous energy she displayed on the basketball courts.

In September 2005, our daughter Shanti was born, followed by Fénoua in 2007. Four girls in the cottage! The construction of the farm was taking shape at the same time as our new family was coming together.

John Seymour, the Champion of Self-Sufficiency

Our teacher, guiding our return to the earth, was John Seymour. His *Self-Sufficiency* book sits permanently on our nightstand.[2] Seymour had an unusual life. In his youth, between the two world wars, he experienced the traditional life of rural England, learned farming, and discovered many crafts that have disappeared today. Seymour's passion for rural life led him on a mission first in England and Europe then around the world to gather people's memories of rural farming life. He put this collective knowledge into practice on successive farms, living self-sufficiently to over ninety years of age. Seymour was a living encyclopedia, the memory of a bygone era. He lived his life to perpetuate simple and effective knowledge that emphasized beautiful, independent living with close ties between man and nature.

Over the years he became a friend. We were inspired by his wonderfully illustrated works to create our fences, our barriers, and a thousand other details that "ring true" and give the impression that our farm has been here forever. Self-sufficiency teaches the essentials of what it takes to live frugally and free in the countryside.

Long ago, most farms had a forge and an anvil. So we decided to give it a try, before building a good one in a few years' time. Heating metal and then shaping it is a difficult but deeply fascinating art. I made a Christmas tree stand that is absolutely indestructible.

A woodstove replaced the oil heat. In winter we rode with our horse and felled trees for firewood — as we do today. Twenty cubic meters (5.5 cords) of wood allows us to heat the house and cook. Certainly, when the cold weather hits, the stove does not always thoroughly heat the poorly insulated cottage. The first one to wake up revives the fire well before dawn. Continuously maintaining the fire and the smell of simmering soup wafting from the stove are satisfactions that blithely outweigh the disadvantages of this form of ecological heating. We wouldn't trade it for the world.

In our busy early years, the farm kept its promises; we rejoiced. It was exhilarating. For me, born in the heart of Paris, it seemed like every fiber of my being was a farmer: I had this job in my bones. As fifty approached, my desire to fully realize this childhood dream, to live off the land, grew.

Perrine and I continued looking for meaning in our lives. We felt a sense of dissatisfaction living comfortably in our little paradise while around the planet an unprecedented ecological and social crisis was erupting. Perrine had already written her first book, *La relaxation en famille* (*Relaxing with Family*) and was teaching relaxation in area elementary schools.[3] But we were both looking for more radical ways to become engaged.

FOUR

The Amazon

Sun, heart of heaven, you must, like a mother, give us your warmth, your light, on our animals, on our corn, our beans, on our herbs, so that they grow, so that we, your children, can eat.

— *Rigoberta MenChú*[1]

For me, ethics is nothing other than Reverence for life.

— *Albert Schweitzer*[2]

*Y*oïwet stretched in his hammock. His wife, Mikilu, was already busy stoking the fire. As the sky brightened above Litany, their children — Pupoli, Pita, and Kuku — left their hammocks one by one and headed down to wash in the river. In the early dawn, brown silhouettes emerged all around; some of them squatting, attending to business, others soaping up on the smooth rocks. Men hid their genitals between their legs, as Wayana modesty requires.

Waking up in the morning with absolutely nothing to feed their three children did not cause a shadow of concern for Yoïwet and Mikilu. Pupoli and his two little sisters headed to the river, line in hand. Yoïwet grabbed his fishing gear, bow, and arrows, and together we pushed his canoe into the water. Mikilu took her katouri, an elegant basket-shaped backpack, and headed off to pick manioc from the garden in the clearing.

At ten o'clock the pot was smoking. The children had caught yayas, curious fish covered with a thick carapace, with a hand line. Yoïwet, for his part, had speared an aimara, a beautiful fish about 60 centimeters (2 feet) long. Mikilu grated the manioc to make a large crêpe. She had also brought back some koumou, heavy clusters of purple fruit picked from a palm tree with a swing of the machete and later turned into a refreshing drink.

Looking back, I cannot help but wonder if a similar morning elsewhere in the world would have been so peaceful. How would we react if we woke up with nothing to eat and no money? For most of us, this never happens, not on a single day of our lives. Yet the fear of going without often worries us.

Local Resources and Individual Skills

From where does the unshakable serenity of the Wayanas come? This lack of fear of the future? The answer is twofold. On the one hand, the fertile nature that surrounds them gives them everything they need to live well, day after day. On the other, they have the skills they need to profit from their surroundings. This explains why the Indians perceive the Amazon forest as a fruitful and generous mother, and why visitors from the developed world often describe it as a hostile jungle, a green hell. We are incompetent to live in this environment so different from ours.

For the Wayanas, almost nothing is purchased, except the products introduced by outsiders that they have come to use — metal tools, weapons, clothes.

Like other first peoples, Amazonian Indians know how to exploit biological resources — wood, plants, fibers, bones, feathers, earth — to make their everyday objects. Boxes, canoes, small furniture, hammocks, bows and arrows, baskets, pottery, and ritual objects are all perfectly adapted to their use, light and strong, nontoxic and biodegradable, yet skillfully decorated. Art combines with usefulness even in the humblest of tools.

Consider these key points that differentiate our modern societies and traditional peoples:

- Food, raw materials, and energy are generally abundant in the immediate environment of tribal communities.
- These resources are collectively owned by the community and are free.
- Money is not necessary.
- Exchanges are based on gift giving or reciprocity.

- Each individual (or each community) has all the knowledge required to meet basic needs through natural resources.
- Everyday objects are functional and beautiful, and the craftsperson takes pleasure in their manufacture.

All this provides a deep sense of security. The first people have few possessions compared with us, but they lack nothing of what they need to ensure their basic needs: housing, food, clothing, tools. Since covering these needs is generally easy, they "work" only a few hours each day — five hours on average for Wayanas (like our prehistoric ancestors, it seems). This gives a different meaning to the notion of work: What they do for work does not define who they are, as in the Western world. Hunting, fishing, growing cassava, and making tools are popular activities. They still have much free time to maintain social relations and celebrate their festivals and rites. Life moves at a very slow pace. When night falls on Antecume Pata, the kololo are set up around fires; men talk about their hunting escapades and their loves, and they speak of spirits. The words are cut by long moments of silence. The serenity is palpable under the stars. Then comes sleep in soft cotton hammocks woven by the women: Couples snuggle, protected from the mosquitoes, the last born sandwiched, skin against skin, between his parents.

At the same time of evening in the West, the main activity might be watching the nightly newscast and the many accompanying advertisements. Silence is anything but normal. Life moves quickly, and we are pushed to live it in a stroboscopic manner, experiencing it in flashes, through commercials, emails, video clips, and the like, all powered with fossil fuel and nuclear energy. Stress builds up; maximum, adrenaline-charged intensity and exhaustion reign. Our inner world seems to crumble; an empty feeling wells up inside that we try to ignore as it's just too painful. Advertising and marketing continually drive our desires and consumption. Yet economic growth stagnates, and, sooner or later, we seek the emergency exit.

My Shopping Cart Has Gone Around the World Five Times

Our modern societies have grown our needs exponentially while distancing us from the places where all our goods are made. The production of food, clothing, and objects has become globalized. The items that fill the carts of Western shoppers travel more than 200,000 kilometers (125,000 miles) on average. We know little about the origin of our everyday products, and even less about how to make them ourselves. We no longer see them as the fruit of human labor; they have become industrial, anonymous, disposable pollutants.

In a consumer society where almost everything can be bought, money has become an absolute necessity. We pay to be born and to die, to breathe clean air, to eat and drink, to get around and sleep, to socialize, possibly even to meet a romantic partner. Yet even as our societies experience greater and greater levels of opulence, we feel more insecure, oppressed. The fear of unemployment is ever present, because losing your salary means it is much more difficult to provide for the basic needs of you and your loved ones. Unemployment becomes a form of social death. The media speak continually of the unemployment crisis, and the crisis in fact is deep. Yet our so-called advanced society doesn't know how to contain this old atavistic terror: the fear of death by deprivation.

What to do? Comparing the lifestyles of ethnic groups encountered around the world with ours in Europe and throughout the "developed" countries, I can only say that our welfare and our sense of security will grow when:

- The origin of the goods we need is as local as possible.
- Our level of autonomy progresses.
- Our skills and our ability to satisfy, by ourselves, a growing part of our vital needs increases.

These findings suggest solutions directly adapted from the observation of tribal communities. After all, their lifestyle was ours until very recently. It is fully possible for us to:

- Surround ourselves with intensely renatured areas, producing most of what we need to live.
- Acquire the skills that will enable us to transform these resources, ourselves, into what we need to sustain us.

This will give us the means to satisfy a growing share of our needs for free — reducing our dependence on outside sources, with the added satisfaction of creating objects that can be both functional and tailored to our own personal use.

The Path of Wellness

To be surrounded by a cocoon of living nature, acquire skills, gain independence, and enjoy plenty of free time to share with family and community: All this increases our well-being far more than a large bank account. Yoïwet, my monkey brother, taught me, without knowing it, the way of happiness. Once our morning meal was finished, he led me to a spot beside the river for my daily lesson on a small flute he made himself from a cariacou bone. That flute, which he played such beautiful melodies on, is now on the table where I write. I have never managed to play it harmoniously. I have great difficulty actually putting into practice the teachings of my barefoot mentors. Constantly, I have to fight against my consumer appetite. I'm more a child of the West than of the Amazon. How I wish, though, to synthesize both worlds.

Certainly, our farm is an attempt to live in the spirit of the Indians, but in this society of ours, the France of the twenty-first century, Perrine and I gradually put into practice a lifestyle that aims to empower us, not cut us off from the culture that we love. The inconsistencies run through us like a fault line. We are often torn.

Yet little by little we see our efforts bearing fruit. By transforming the space around us into a kind of more intensely alive bubble, we feel nourished and protected by a matrix, like a baby in the womb. The gardens and trees around us in concentric circles form a kind of generous placenta.

Sometimes at night, when the valley basks in a warm light, I can perceive my surroundings in the same way that Yoïwet spoke of Mother Earth. I hear her heart beating, the blood pulsing through her veins. Her wind touches my skin like breath, her rain cleanses me, her sun energy animates us — the plants, animals, and farmers who share this valley.

FIVE

We Are What We Eat

The body is a sacred temple that needs to be loved with passion, sweetness, tenderness. If you can say you love your body completely, then you can be true when you say, "Love your neighbor as yourself." For to love one's body is to love the Earth. To destroy one's body, is to destroy the Earth. To make one's body beautiful is to embellish Earth.

— Don Marcellino, Huichol healer[1]

Located just outside the cottage, our garden slopes gently toward the river. Less than a meter separates the kitchen door from the aromatic herb border. We cohabit for months with the fruits and vegetables that feed us before we consume them. Believe me, this completely changes our relationship with food.

Take the cherry tree mentioned at the beginning of this book. It grows, the trunk thick like a man, just outside our windows. In April it dazzles us with an immaculate flowering for ten days. Then we watch the fruit set, hoping that a late frost does not spoil the harvest. Cherries form, grow, and begin to color. Birds bring joy to the heart, fortunately in the upper branches, out of our reach. Then comes the time for cherries, finally ripe. The generous tree offers us much more than we can consume; fresh cherries and clafoutis grace the table for two full weeks. We remove the pits from the extra cherries to make syrups and jams, and the remainder is dropped into brandy. All these little details give a good flavor to life and really have nothing to do with buying a kilo of cherries anonymously at the grocery store. Our tree brings us pleasure for several months. An aesthetic and gustatory pleasure, free and annually renewed. If only everyone could have a cherry tree in their life.

To extend that pleasure, we have planted several dozen different varieties of cherry trees, the earliest to the latest bloomers, producing a wide
variety of fruit colors and unique flavors. We did the same for the plums,
pears, peachs, apples, apricots, figs, and nashi pears: No less than five hundred varieties of fruit trees now populate the orchards at La Ferme du Bec
Hellouin. To which is added over a good hundred varieties of berries.

Salads and Leeks for Neighbors

A small greenhouse built at the end of our house allows us to plant our first
seeds as early as January. In March the seedlings are transplanted. We
monitor their growth with a watchful eye, waiting for the leafy vegetation
to emerge. Nothing compares to the flavor of a salad picked minutes
before the meal. There is no doubt that the vital energy it contains is
infinitely stronger than that of its cousin harvested a week earlier and
transported to warehouses in refrigerated trucks to land in a supermarket,
the leaves at half-mast in a fridge before being ingested. A fresh-picked
salad is alive; the other is dead. The tragedy is that virtually all our food is
dead today. Therefore, learning to grow our own food is a source as much
of health as of pleasure.

I talk a lot about fun on these pages. A farm is a place of generosity,
sharing, conviviality. A land that nourishes the body and rejuvenates the
heart. Perrine and I have reflected much on the act of eating, over the years.
Initially, our sensibilities were diverse. Perrine, former high-level sports
player, having long suffered from stomach pains, was interested at a young
age with the relationship between food and health. And as a sailor, I had
tasted the most unlikely foods all over the planet. Then we both changed
careers and became psychotherapists. Passionate about a holistic vision of
the human being, we were trained in relaxation therapy, therapeutic massage, yoga, and ayurveda.

Our profession and our early years on the farm made us far more
aware of the vital importance of diet. We are what we eat because our
body is made with molecules that once made up our food. Is this not an
invitation to respect? As Huichol healer Don Marcellino said, taking care
of your body means taking care of the earth, since your body is a reflection
of the cosmos.

Do You Want to Become a Tagada Strawberry?

When groups of children come to visit the farm, I help them understand this concept with an image: "If you eat only Tagada strawberries [a favorite French candy], your body would become big, pink, and soft like giant strawberries," I tell them. They generally laugh, but it's a forced laugh. Researchers have found traces of more than two hundred different toxic substances in the umbilical cord of the human fetus.[2]

If we eat industrial products, refined, polluted with pesticides and herbicides, laden with preservatives, additives, and bad fats, how can we hope to be healthy? Conversely, eating organic food — alive, and ideally produced locally, or better yet by ourselves — provides our body with essential nutrients, taste, and more pleasure.

Consider the four million years during which our human ancestors gradually evolved. They were surrounded by nature, living only on wild foods: leaves, fruits, berries, roots — and a little meat and fish occasionally, rather late in the evolutionary process. Our bodies are made for this natural food, and totally unprepared for the "modern" diet, industrial and chemical, saturated in grease and sugar, too rich in meat, cereal, and dairy.

Eating with a Conscience

When we eat today, we no longer listen to the needs of our body; we listen to commercials. But do shareholders of industrial agriculture corporations really care about our health? The explosion of food allergies and chronic diseases like diabetes, obesity, and cancer attests, unfortunately, to the food and health drift in which our so-called advanced societies are stuck. It's small consolation that all these diseases of modern civilization raise the GDP.

Is it reasonable to let corporations, forced into a logic of profit in the short term and competition for market share, decide how our bodies are formed? Are the industrial food corporations competent at replacing natural food, which fed us for almost all of our history on the earth, by food containing several thousand different synthetic molecules whose effects combined and long-term have not been tested?

When we know the suffering experienced by a cancer patient and the magnitude of the number of people affected (more than fourteen million new cases are diagnosed each year around the world, with treatment thought to cost hundreds of billions of dollars[3]), it is surprising that we do not collectively choose to prioritize agriculture and food that is truly natural.

Twenty-four hundred years ago, Hippocrates advised, "Let your food be your medicine." More and more of us are rediscovering the truth behind this simple statement. All organic gardeners contribute to restoring their health and that of their relatives thanks to the quality of the harvest.

Nourishing Ourselves Like We Make Love

Feeding ourselves is an act as intimate as lovemaking. When you make love, the fusion is complete. But the intimacy of the bodies goes far beyond the physical: The energy of your lover sinks deep into your energy. This is why the sexual act, depending on how it is experienced, can be a source of immense benefits for deep disorders. The same synergy happens with every meal. The molecules ingested join up with each of the billions of cells that constitute us.

Ingesting polluted industrial food is a form of assault for our bodies. An assault to life. Giving these foods to our children can be seen as slow assassination. We all want the best for our children, right? Becoming more aware of the role of diet can help us make the best choices for them. Healthy diets should be part of our basic education, as should stress management and nonviolent conflict resolution. But the important things are not taught in school. We note at the farm that many young adults don't know how to cook vegetables, or even know the differences among onions, shallots, and garlic.

Eating is a sacred act that should be surrounded by the greatest respect. Food, like sex, should be a matter of love.

Vital Energy

Beyond the chemical-molecular composition of our food exists a more difficult-to-understand, though just as real, dimension of energy. Our Western mentalities, governed by analytical reasoning, are still unfamiliar with the concept of vital energy, unlike the Asians who call this energy prana in India, qi or ch'i in China, and ki in Japan. According to them, a fruit or vegetable grown naturally, gorged with sun, consumed as soon as possible after harvest, is full of prana. But the more cultivation methods depart from the laws of nature, the longer the product is stored after harvest, the less it contains prana. Similarly, cooking, especially in a microwave oven, refining, processing, freezing, and any other handling that removes food from its natural state causes it to lose all or part of its vital energy.

Asians go farther and recognize that the state of mind of both the gardener and the cook influences the quality of the food. A product grown with care and cooked with love contains positive energy, beneficial to the health of those who consume it. Our common wisdom recognizes "a good dish cooked with love." We intuitively feel that specialness that comes from attentive, careful cultivation and preparation.

Parents, empty your freezer, put the Tagada strawberries, prepared dishes, packets of chips, and chemical cakes in the trash, and take the time to cook fresh, seasonal food for your children. This is one of the greatest gifts you can give them.

Natural Food, Foundation of a Lifestyle

We all realize that modern lifestyles, increasingly urban, deprive our bodies of our basic needs. Our lives are artificialized. Nobody wins when a significant part of oneself is not connected to the biosphere that has fed us throughout our evolution.

What joy, though, there is in taking care of one's health through diet. In the spirit of the Slow Food movement, which is gradually winning over the planet, food may be the first link in a process that will gradually establish a true art of living, based on listening and respect for self, others, and life. Favoring local food allows us to reconnect with the earth through local small farmers, who can become a real interface between urban life and nature. From these fruitful exchanges comes a shared feeling of gratitude. Everyone wins in this exchange: Did you know that gratitude is the sentiment most favorable to health?

Finally, reclaiming our choice of food is a political act, a way to move our society in the direction of our aspirations. Giving power back to the small farmer and local agriculture means taking care of the planet and ourselves.

Nearly three years after creating our small family farm, our interest in the therapy profession evolved. It seemed to us that feeding people with healthy, totally organic products, grown with love and respect, was an essential task. So in the summer of 2006, we decided to become farmers "for real" — that is to say, to live off the sale of our production. Saint-Exupéry maintained that the three best jobs are aviator, author, and farmer. We were about to get the chance to experiement with one of the best jobs in the world.

SIX

Draw Mc a Farm

Do as much as possible for and as little as possible against.
— Gilles Clément[1]

In October 2006, the organic Ferme du Bec Hellouin was officially born. We took on the role of farmers with absolute naïveté. Neither Perrine nor I had ever stepped foot in an agricultural school. We had never even visited a commercial organic farm! We imagined that our farming business would be an extension of our previous years' family garden. The gardens would just be bigger. We did not measure the gulf between achieving family food autonomy and making a living from selling our agricultural products.

We continued working on the site we'd begun developing in 2004. Our business was growing almost exclusively out of a 1.8-hectare (4.4-acre) area that surrounds our cottage at the end of the Bec valley. We chose to put down roots in the rural tradition. Over the years, we added buildings all made by local artisans using local, recycled materials: stone, wood, bricks, tile, clay. I admit to a passion for old stones, and I took enormous pleasure designing and building these structures. The store at the entrance of the farm and some shelters were all planned and the sites were prepared before we attacked, from 2009 to 2011, the huge construction site that became our ecocenter. The construction had no apparent order. Rather, it followed a carefully studied disorder, in the orchard that extends north of the river.

A Farm, Like a Painting

Like John Seymour, Lucien Pouëdras unknowingly influenced the genesis of our farm. Born into a family of small farmers, he became an artist late in life, painting the rural Brittany of his childhood, a country that remained very traditional coming out of the war, in the 1950s.[2] His works from that era are imbued with a poetic charge that touched me deeply. I walked for hours in his paintings. They depict a rural wooded landscape, interlacing fields and farms, valleys and tree lines, woods, streams, fields, and pastures. A landscape that shows the beauty of the shrubs, coppices, and fruit trees. This Breton countryside is intensely inhabited: All men, women, and horses are at work; children participate in the work of adults or hunt fox; an old man is bent over in the garden, a grandmother is milking her cow.

The paintings resonated deeply with me because they evoked similar scenes from my childhood. We spent many of our holidays in Vendée at our aunt Marie Therese's house. During the 1960s, the Vendée had not yet been damaged by bulldozers. We would set out on bicycles to fish in the Lay creek. Small meadows carpeted with buttercups were pure wonders. The Paris child I was passed some of the finest moments of his life along that creek, watching the sunlight play across running water, observing the small, minnow-like gardons swimming in the water.

My paternal grandfather had a house in Dorlisheim, a village in Alsace. Whenever I heard horse hooves there, I ran toward the fence at the far end of the garden, climbing to watch as a heavy horse passed by, pulling its cart. It was the neighboring farmer on his way to work in the vineyards, which blanketed the whole width of the hillside. One day, I jumped the wall and went to explore the vineyards. I climbed along the narrow pathways in the middle of rows of vines, hands sticky with the juice of peaches picked along the way. At the top of the hill, the view stretched far away across a rural landscape drowned in the summer sun. Workhorses sometimes appeared in the vineyards. Nestled in the valley was the village, and its massive farms.

Much later, my grandfather's house was sold. I returned one last time, struggling to find the village of my childhood in the maze of expressways. At the edge of the park I climbed the wall. No more horse, of course, just a wide paved road and beige houses smartly lined up in front of their sidewalk in a subdivision. Dorlisheim today is almost fully connected with the suburbs of the city of Strasbourg. Farmland is disappearing rapidly in France, drowning under tar and concrete. We lose the equivalent of one French department every ten years.

I have always been pulled to these memories. They have influenced our course. Perrine and I began, without being aware yet, the construction of a very unusual type of farm in the world today: a microfarm where everything is done in sync with the rhythm of life.

A Mosaic of Ecosystems

La Ferme du Bec Hellouin is designed like a painting — a sort of living, three-dimensional one, with vibrant colors and sunlight, wind, river, and clouds. As in a Pouëdras scene, there are no straight lines, but many sensuous curves interlacing orchards, pastures, gardens, and ponds. And everywhere, there are fruit trees, by the hundreds.

Our first goal has been achieved: This tiny territory is home to a wide diversity of intimately intertwined species. We created an edible landscape. It surrounds, nourishes, and protects us. Enveloped by our hedges and our trees, we are immersed in a life that feels very secure. Each year, as this microcosm begins to mature, I find more of that feeling of well-being that I experienced in tribal communities. A synthesis occurs gradually. We set out with our children, basket in hand, and pick berries, fruits, and flowers randomly as we walk, just as the canoe in the Amazon filled up with good things as it navigated the river. It is as if there is a hint of the virgin forest in our Normandy countryside.

Giving sway to beauty seemed an obvious choice. Beauty is a food as essential as bread. Without it, the soul atrophies. Traditional farms from every continent were beautiful, simple expressions of their landscape. Today, they too often resemble factories. Burdened by the workload and weight of loans, farmers resign from their role as guardians of the landscape. We are the only animal species to uglify, massively, the surface of the earth.

Our approach was essentially intuitive: We spent a lot of time observing and listening to what resonated with us before altering the landscape. Later, as we received visits from many agronomists, we came to understand that this highly diversified agroecosystem, created with aesthetics as a top priority, is also highly productive because of the exchanges among all those little, interrelated circles of trees and crops, allowing them to get the most out of the services rendered by their ecosystems.[3] Without knowing it yet, we instinctively implemented many of the principles of agroecology and permaculture. Productivity was somehow given to us as a bonus, a good surprise revealed over the years.

Our visitors feel good about this human-scale farm that so closely resembles those depicted in our children's books. (We all have a small farm trotting in our imagination.) Almost invariably they tell us, "This is paradise!" Do they know that the ancient Persian *pairidaeza*, the root of *paradise*, means "garden"? Our collective unconscious remains marked by images of the rural life of yesteryear, back when humans, plants, and animals all lived very close to one another. This is what we need to survive in a concrete world that does not meet our most profound needs.

In the summer of 2006, we had no idea where our adventure would take us. I felt the same pinch in my stomach as I felt before embarking on an ocean crossing. Our goal then was simply to live happily and peacefully in our gardens and feed a few dozen families with the best possible products.

SEVEN

New Farmers

The economic purpose is not the secretion of more but the advent of better.

— *Jean-Philippe Barde and Christian Garnier*[1]

To make a goal of comfort or happiness has never appealed to me; a system of ethics built on this basis would be sufficient only for a herd of cattle.

— *Albert Einstein*[2]

*J*n our first season as market gardeners, 2007, the whole farm was cultivated with animal traction, thanks to the efficient collaboration of Winik, our Mérens horse.

Being neophyte farmers, we made some radical choices. Under no circumstances were we going to lay down plastic ground cover (widely used in organic gardening to limit weed growth) and pollute the landscape. So creating our gardens in what had been, just a few months before, a pasture, we continually fought against the tendency of Normandy grass to resume its place with vigor.

A Little Bit of Earth and Many Stones

We also installed our first market gardens at the very bottom of the valley. It was the only available earth we had at the time and we did not ask questions. But locals would tell us, "You can't farm the valley!" We began to understand why when we felt our tools scraping against rocks as soon as we sank them a few inches into the soil. We realized that there was only a thin layer of topsoil—depending on the location, from 10 to 20 centimeters (4–8 inches) thick. Below that lay a meters-thick, compact bed of flint. We consoled ourselves at first by thinking that while we had little soil, it was black and fertile. Soil analysis soon disabused us of this notion: The soil was very, very low on the fertility scale. It contained a lot of organic material, but that material was "blocked," with little available for plants, particularly because of excess calcium. The engineer from the Chamber of Agriculture, which conducted the analysis, told us that we had one of the most unfit soils of the region for cultivation. The crops in our township were all located on the fertile plateaus, while the valleys were vested in animal breeding. An archaeologist topped off the bad news by telling us that since Neolithic times, these valley floors had never been cultivated.

The alliance of our abysmal inexperience and our poor soil quality strongly held back our chances of success. The good results we would get in a few years was not therefore explained by exceptionally fertile earth or farming knowledge, but by the permaculture approach.

Between Enthusiasm and Depression: The Quest for a Business Model

We also wanted to market our products at the farm and decided to build, with our closest neighbor, a shop, constructed mostly from trees from our own woods. We believed, again naively, that our customers would be delighted to do almost all their shopping with us. So we set our sights on diversifing our product line. In addition to vegetables and fruits, we produced cider, apple juice, apple cider vinegar, bread, syrups and jams, and satchels of aromatic and medicinal plants. In our zeal to do more, we decided offer the excellent dairy products from the nearby Pincheloup farm, and have a corner for organic products that we don't produce.

In theory, it was all fine. The farm's "résumé" was laden with the trendiest features—a small-scale family and market garden structure for processing the crops on the farm and transforming them into products to create added value and sell locally. In practice, though, we ran up against two major obstacles.

The first: Each new product required investments in equipment and a suitable locale (we dug a cider cellar, for instance). It also required the acquisition of skills — not only to transform crops into products but also to comply with health and labeling regulations for every product. The certification for organic agriculture added an extra level of complexity. Each recipe and each label, for example, had to be validated by Ecocert, our organic certifier. All this took time and money.

We underestimated how many different directions diversifying would take us in, and the corresponding fatigue. We ran endlessly. Perrine was admirable, present on all fronts, making jams until late at night between feedings. Without her inexhaustible energy, La Ferme du Bec Hellouin would have soon failed. Today, we advise those who come to learn at the farm to diversify in stages, not beginning a new product until the previous one is mastered. We also suggest future farmers not give birth to babies and simultaneously start a farm — unless they want to forget what sleep means for a few years. At La Ferme du Bec Hellouin, we ended up keeping a wide variety of products, but the price was really high and it has taken almost ten years to get a return on our investment.

The second major obstacle was our location. Despite our proximity to the abbey of Le Bec Hellouin, in a village ranked among the most beautiful in France, we did not have enough local customers for all of our products. In our Upper Normandy countryside, many people cultivate their own vegetable gardens and the prevailing mind-set is not that open to organic agriculture. Upper Normandy was, in 2006, the bottom of the barrel for organic agricuture, with only 0.4 percent of the cropland farmed organically. The rest, a full 99.6 percent, was farmed conventionally. This was also the region of France that used the most pesticides. Rates of cancers and obesity were among the highest in France. And yet we were told from time to time, "Organic food makes me ill." Quickly, we resolved to sell a portion of our production in the surrounding small cities (Bernay, Bourgtheroulde), but especially in Rouen and Paris.

A Solitary Journey

Another difficulty of the early years was our isolation. The nearest organic farm was 20 kilometers (12 miles) away, and eight years later that has not changed. No market gardener is there to share experience. We accumulate blunders. There is a regional association of organic farmers in Upper

Normandy, with a friendly, competent, and committed team. But a lack of resources and personnel meant that it was possible to have only one technician visit the farm per year. This lack of support slowed our progress. It was clear that despite the facade of ads, successive governments had not implemented the human and financial resources for the real development of organic agriculture in our country.

On top of that, we were just too unconventional in this region of big agriculture, where the tractor is king. I remember the first visit by Aurélie, our market gardening adviser from the association. When seeing our crops, she laughed with surprise: "But this is just a big garden!" I began to lose confidence, a feeling that was accentuated when the technician pointed to a row of carrots saying, "What are you going to produce with that? Not much!" She lavished us with advice, and we asked her if we were completely off our rockers, if we were setting ourselves up for utter failure. Visits to other organic market farms later revealed that our colleagues have very different types of practices: large open fields, many greenhouses, garages full of impressive tractors, and a fairly pervasive use of plastic ground cover.

We just had a large garden. It was perhaps not viable. Moreover, even our mechanized colleagues were struggling to get by. However, we did not envision practicing our profession in a mechanized manner. Motorized vehicles and I share a mutual dislike. Every time I buy a lawn mower, chain saw, hedge trimmer, or vegetable mill, the tool breaks down in no time. I think when motors see me coming, they perceive my trepidation and enjoy playing tricks on me. I am much more comfortable with my horse. He makes no noise, offers strength and dung, and becomes more and more of an accomplice with each passing year.

Later, François Léger, an agronomist with whom we launched a study on the original crop system implemented at La Ferme du Bec Hellouin, would tell us, "Your farm is not the reduction of a mechanized farm, but an extension of a large garden." Today we acknowledge this without blushing. The productivity achieved over the years is partly explained by our initial choices.

The flavor of our vegetables, though, was a positive. We were always pleasantly surprised when customers praised it, and told us that it reminded them of the long-lost taste of vegetables from their grandfather's garden. We still have a lot of progress to make to improve our production. But today we have the satisfaction of working with several renowned chefs from starred restaurants, who confirm the impressions of our first customers.

An Adventure at Risk

The economic results from our first year of market gardening were disappointingly low: We both worked like crazy to bring in an absolutely pathetic 12,000 euros. Perrine roundly remarked that in her previous career, she earned as much in a couple of months. Given the costs involved, we were heavily entrenched in the deficit column.

In the interest of transparency, it should be noted that we held on during these early years due to our previous professional lives, which gave us enough starting capital to finance the initial investments for the farm without going into debt. And I continued to receive royalties from the distribution of my documentaries — on marine life and native cultures — that have been sold in many countries for some years. Without these revenues, we would have been forced to give up our work as farmers and take better-paying jobs to support the family. Our prior professional experience had in some ways "sponsored" our early farm system, so badly prepared, and bought us the time to find solutions to our problems. We are aware that this particular situation is unique, and highly recommend that future farmers plan every aspect of their start-ups carefully, without neglecting any details. This would prevent many setbacks. When our savings ran out and we plunged into debt, we chose to continue at all costs. Renouncing is not in our character. But the cost was heavy. We had a few years of great anguish.

The Old Man and the Turkey

Despite the setbacks, though, we were awarded the ecotourism trophy for Upper Normandy less than a year after we opened the farm to the public, and without having solicited it. We were happy to see smiles on the faces of almost every visitor. They kept coming back. Many of our customers from this first year are still loyal eight years later, which is the best reward. They appreciate the bread out of the oven and remain for hours with us when we press cider. Some, like Bruno and Martine, Minnie and Romain, Bernard and Nicole, Evelyn, Karine and Sebastien, Virginie and Pascal, have become friends and help us stand firm in difficult times. They accept the imperfection of our produce and are not too alarmed about slug holes in the leaves. They encourage us, week after week. A former forester, Bernard has volunteered to trim hedges and mow for several years, always with good humor.

The social success of the farm was inversely proportional to its profit-and-loss statement. We realized that our aspirations were not so different from those of the majority of our contemporaries. We followed our hearts when creating this space, and visitors were touched by it. One day, I saw an old man in a suit and tie, very dignified, sitting by the river with our turkey on his knees. The image struck me because that bird was huge. We learned that a farm can produce much more than food for the body: It also feeds the emotions, the souls of our visitors, and meets their growing need for reconnecting with nature. As new farmers we became a kind of interface between our contemporaries, predominantly urban, and the rural world — a world increasingly marginalized by French society, which has its codes and its culture, and remains difficult to penetrate for the uninitiated. For centuries, many rural farmers have suffered the contempt of those they feed, and sometimes respond with legitimate suspicion.

This first year of market gardening had upended our dreams. Where was the quiet and bucolic life to which we aspired?

Rock-Bottom Spirits

The 2008 season ended with our morale at half-mast. Given our success with tourists visiting the farm, we had decided to develop a form of agri-tourism. We turned our attention to the farm's welcome center, hiring three people for the season and opening to visitors six days a week. That first year we registered thirty-five hundred paid admissions. We organized a musical tour of the farm twice a week. The result could have been encouraging, but for us it was a nightmare. Having so many visitors under our windows became intrusive, and family life was turned upside down. We were all the more vulnerable with Fénoua, our fourth daughter, born during winter, who had not yet decided to let her parents sleep at night.

On the crop front, we had installed a 580-square-meter greenhouse and production increased significantly, but the money we earned all went to pay employees. For the second year running, we could not pay ourselves and we were heavily in debt. Then we hit our limits in terms of workload.

Our life as organic farmers was nothing like our initial dream. We never imagined it would be so hard — working so much at the limit of our strength, trapped in an endless race every day, never a moment to contemplate what we created. Our savings were running out, and we were heading for a deep debt spiral. We both agree that the atmosphere in the cottage then was not exactly bucolic. Often I interrogated myself: Had I done the right thing engaging Perrine in this adventure? Would our family explode under such pressure? Yet when I looked deep inside, my little inner voice would tell me, "Go on, your path is right." And I persevered despite the doubts because I am more stubborn than our donkey.

The end of the season raised technical concerns, too. We both questioned the merits of animal traction. Certainly, working with animals was marvelous, but plowing with the horse meant we had 75 centimeters (30 inches) between each row of vegetables — a small desert of naked earth, grilled by the sun or washed out by the rain. Perrine reflected even further: She was not comfortable with the horse-drawn metal tools disemboweling the earth. Even if they were pulled by a horse, it was not very natural.

We journeyed toward another approach.

EIGHT

Discovering Permaculture

Perhaps we seek to recreate the Garden of Eden, and why not? We believe that a low-energy, high-yield agriculture is a possible aim for the whole world, and that it needs only human energy and intellect to achieve this.

— *Bill Mollison and David Holmgren* [1]

*I*n October 2008, a friend sent us an article that introduced us to a new concept: permaculture. It was possible, we read, to base the design of our farm on natural ecosystems and work with nature, not against it. Permaculture was largely unheard of in France, with few practitioners. We wanted to know more, but there was no French literature on permaculture at the time. So Perrine ordered books in English, and launched into extensive research. What we discovered impassioned us.

This new method was based on observing the functions of natural eco-systems — which are sustainable, self-reliant, and resilient — and using them as models to create more sustainable and autonomous human habi-tats. The central idea of permaculture is to create a network of beneficial relationships among all the components of a system, so those components can interact. It is above all a conceptual approach. Permaculture design is primarily a process of observation and reflection to correctly position the elements of a system among one another, so they can interact. A carefully designed system will be efficient with inputs and energy, requiring less effort to operate while being more productive.

But permaculture is far more than a set of growing techniques: It is a flexible and pragmatic framework, an invitation to put ourselves in the school of nature and creatively transpose its teachings to all sectors of our life. Permaculture design allows us to coherently integrate technology borrowed from various fields, such as agroecology, sustainable building, and renewable energy, as well as innovative approaches to governance (from nonviolent communication to sociocracy[2]) and sustainable economics (from the circular economy to local exchange trading systems[3]).

First and foremost, permaculture is born from common sense, a way to consider the system as a whole. A child can understand its essence intuitively. Many of us are permaculturists without knowing it. But awareness of these key concepts, gained from observing ecosystems, allows us to gain efficiency and consistency:

- In nature, everything is connected.
- Ecosystems function in a loop.
- Each element helps and receives help from the others.
- The waste from one is a resource for the other.
- Everything is recycled.
- Each important function is performed by several elements, and each element potentially performs several functions.
- The whole is more than the sum of the parts.
- Each ecosystem works largely independently and makes a contribution to the whole biosphere.

Walking in the Direction of Life

Permaculture has influenced much of how we think about our farm — as part of our local economy, as an experiment for a post-carbon future — but our primary focus at La Ferme du Bec Hellouin has been on how it applies to agriculture at the home garden or market farm scale, and much of our work now revolves around developing and sharing new insights from research conducted at La Ferme du Bec Hellouin and other sites.

Forests are a powerful source of inspiration for permaculture. Without any human intervention and without inputs, they typically produce two times more biomass per hectare per year than our cultivated agricultural systems. One hectare of chestnuts produces as much vegetable protein as a hectare of wheat, the latter requiring much human intervention, fossil fuel, and other inputs. But a large share of the forest's biomass (certain leaves and undergrowth plants) is not directly useful to humans whereas in the wheat field, everything is valued — straw as well as grain. How to combine the autonomy of the forest and the food and resource production of a wheat field?

Bill Mollison and David Holmgren, the two Australians who coined the term *permaculture* in 1978, have sought ways to apply the forces that make natural ecosystems so productive and independent to our human settlements. For example, observing the forest gardens of many people living in tropical regions, they imagined a new type of forest garden for colder regions in which all (or almost all) plants are edible. The edible forest garden is an imitation of the wild forest, adapted to the needs of humans.

Growing in Raised Beds

During our initial research, our attention was drawn by a system of growing that has been around for thousands of years, rescued from obscurity by permaculture — growing in permanent beds. This approach is based on a simple observation: In nature, the soil is never plowed, never worked. In addition, it is generally always covered by a layer of decomposing matter. By creating permanent beds, we avoid destroying soil fertility with mechanized equipment or digging. The soil organisms — worms, bacteria, fungi, and algae — will be able to prosper and naturally improve the soil's structure and fertility. If the ground is also covered by compost or mulch, fertile soil elements no longer leach away, weeding is reduced, moisture reserves are protected from evaporation, the first centimeters of the soil are not sterilized by the action of the sun, and as this material decomposes it composts in place.

The act of creating a bed of good topsoil has many other advantages: The thickness of the humus layer is increased, and plants have more to eat. It warms up and loses its excess water more quickly at the end of winter. The ergonomics of the work is improved: Gardeners circulate through the beds exclusively in permanent walkways, the soil is never compacted, and you don't have to bend down as far to reach the plants. There are also more niches for beneficial insects.

According to British permaculture expert Patrick Whitefield and biointensive agriculture leader John Jeavons, cultivation in beds was practiced in China four thousand years ago, and in South America over three thousand years ago; the Greeks were doing it two thousand years ago.[4] It is still practiced in different parts of the world — in Melanesia, for example, for root crops. In traditional home gardens, raised beds were also a form of permanent beds.[5] In my grandfather's garden in Alsace, I remember seeing raised beds supported by a brick formwork, which would be covered in winter by glass frames; such structures are both aesthetic and highly productive. In our quest to gather information on permaculture, we quickly stumbled on the website for Au Petit Colibri, the farm of Richard Wallner, also in France. Unable to expand his vegetable production because of administrative red tape, Richard devotes his energy to research and the dissemination of information.[6] He sent us his DVDs. One of them depicted the garden of Emilia Hazelip. Now deceased, Emilia was a pioneer of permaculture in France. Her garden was inspired by mandalas — centered geometric shapes that promote calm and concentration and fill a spiritual function, seen often in India. There is energy in this design, which we discovered again on satellite photographs of gardens dating back thirty-two hundred years in Peru's Lake Titicaca region.

The Mandala Garden

Despite the exhaustion that came as our season drew to a close in November 2008, the designs spurred great motivation. Without waiting, we began constructing our own mandala garden. The idea of gracing our landscape with a design that was thousands of years old spoke deeply to us. Once again, our market garden profession had given us the opportunity to merge beauty and food production — to cross, in our Normandy valley, influences from ancient times and other cultures, from India to pre-Inca America. We transformed an 800-square-meter space, previously worked with animal traction. I was so taken by the project that I woke up early, pacing until dawn, and found myself picking at dirt that was still frozen in the first few rays of daylight. The site was created entirely by hand, with the efficient help of Jean-Claude Bellencontre, who joined the team and has become, over the years, one of the pillars of the farm.

We completed the mandala garden in February. It wasn't perfect, but its design as well as our manner of growing in its beds would be improved from year to year.[7] Yet even in 2009, it was incomparably more productive than the same-sized space cultivated with animal traction — by a factor of five, at least. We eventually grew a row of vegetables every 75 centimeters (30 inches), with a bare strip between rows. Today the beds are beautiful and deep, about 1.2 meters (4 feet) wide, fully covered by crops. The design has been an enormous help in controlling native plants (I prefer this term over *weeds*, which does not pay homage to these courageous plants), and the agronomic conditions in which plants grow were greatly improved. Moreover, the visual impact of the mandala garden is strong and contributes to the beauty of the farm, a benefit to both workers and visitors. In the years since we created it, we have passed along the technique to many trainees who come to learn at our ecocenter, and they report similarly positive results.

Island Gardens

By the time spring 2009 rolled around, Perrine, a prolific Internet researcher, had spent the winter studying permaculture. The Internet is one of our most valuable tools on the farm. What good fortune to be able to watch, in the evening after a day of work in the gardens, films shot in Australia, California, Nepal, and Africa. Never before has the disemmination of good practices happened so quickly, which fosters real hope for saving the planet. That spring, Perrine left (with difficulty) the family and the farm, to attend a two-week certified course on permaculture in England.[8] We also decided to transform the western part of the farm — a major project. The land there, a few thousand square meters, was very poor. The topsoil was at its thinnest, and the grass was just good enough to feed some skinny sheep. It would become, over the next several years, a lush oasis.

The design we applied there was bolder than anything we'd tried before. It was based on a long initial period of observation. For several years, we had been thinking it through and working at small pieces, but our growing familiarity with permaculture gave us the conceptual tools necessary to go much farther. The first thing we addressed was water, since the Bec river flows along the meadow there. We had already dug, two years earlier, our first small pond. The presence of water in a landscape gives it additional beauty: a mirror reflecting the sky; a stop for herons, egrets, ducks, and snipe; a permanent habitat for waterfowl. We really wanted to push this work farther, to develop the land–water interface, which is, in all latitudes, highly productive.

I have long been fascinated by vegetable gardens located just alongside water — like the hortillonages of Amiens, vast floating market gardens located in waterways north of Paris and cultivated since Roman times. The term *hortillonnage* refers to marsh gardening, as wetlands provide the abundantly fertile land and water required for food crops.

We began by digging, with permission from the mayor of Le Bec Hellouin, a network of small ponds defining two island gardens. Then, to compensate for the absence of soil, a fellow farmer brought compost from Le Bec Hellouin horse club and spread it in a generous layer; there was enough to create mounds on both islands. Charles Barbot, the club owner, still allows us to collect compost from his manure pile. This transfer of organic matter illustrates the spirit of permaculture: Any waste from an activity, if it is not recycled within the system, becomes a pollutant on the outside. Waste reused from a local riding club becomes a resource for the farm.[9] Ever since, when creating new gardens in areas that are rocky and unsuitable for crops, we call on the horse club for compost. It only takes one application and a few hours of time to build a foundation on which we can then gently, patiently, and lovingly create good earth for growing vegetables. Not everyone has locally available manure, but it can be replaced with green manure, or you can plant trees and perennial crops to help fix nitrogren in poor soil.

The Forest Garden

March came. The two islands were created, the beds formed. This normally would have been the time to prepare for the gardening season. But Perrine discovered, through her studies, the innovative concept discussed earlier in this chapter: edible forest gardens. A forest that can be eaten? It didn't take long for our imagination to ignite. The wonderful concept of a forest garden spoke to us profoundly. Despite the enduring fatigue and the fact that spring was upon us, we embarked on the creation of our forest garden. It now occupies an area of 1,200 square meters and runs in a horseshoe shape around the ponds and islands, on the side of the prevailing winds.

The forest garden became Perrine's kingdom. She spent her days and nights in search of suitable plants. In France, potential suppliers can be counted on the fingers of one hand. Ultimately, a hundred varieties of fruit trees, shrubs, berries, nuts, nitrogen-fixing plants, ground covers, and edible vines would be planted, some of them originating at the Martin Crawford Agroforesty Research Trust, in England.[10]

In April we were forging ahead. There is a principle of permaculture we have not mastered, the one about doing things slowly! We were more than a little proud of the result, though, and it was fascinating to see how, in just two months, the mediocre field had transformed. Now there were four different areas: the forest garden; the island gardens and ponds that the forest garden's curves provide generous shelter for; and a small willow orchard that serves as a paddock for Winik, our draft horse, Alice, our donkey, and the kids' ponies. These four areas were closely intertwined, with numerous edges. The land–water interface was maximized, running several hundred meters.

The Whole Is More than the Sum of Its Parts

By summer the crops on the big island were truly surprising, and today this island has become the most productive area of the farm. Over the years the forest garden grew in magnitude. Lacking experience, we mismanaged the ground cover early on, and nettles quickly took over. But we learned to see them as a resource: Rich in minerals and nitrogen, they make an excellent mulch for beds. The forest garden, too, has continued to flourish.

All this required much hard work for two months that spring, with the help of an excavator, tractor, and imported compost. But then we could enjoy the results: We could ease up, work only by hand — no more mechanized craft entered this space — and observe the way that life emerged in the new agroecosystems. The ponds became populated with frogs and aquatic plants. Curiously, trout, sticklebacks, and even white-clawed crayfish arrived, seemingly spontaneously. The dazzling kingfisher became a daily visitor. A mallard duck and her twelve ducklings swam in the small creek that separates the forest garden from the islands.

Years passed, and this space became more and more beautiful, fruitful, and productive. We have been rewarded beyond anything we could have imagined and our farm has become living proof of the permaculture adage, "The whole is more than the sum of its parts." These small ecosystems interact. Organic matter circulates. The resources are numerous. Reeds grow rapidly in the ponds, forming an abundant biomass, edible for both humans and animals. The muddy pond bottoms concentrate fertility, which in turn fertilizes the beds. Nettles, comfrey, and burdock are mineral pumps that grow in abundance; we have only to deposit the mowed mulch on the beds. Trimmings from the forest garden and the willow coppice around the ponds are turned into wood chips. Horse manure feeds the compost pile. Branches of ash trees that grow along the river are an excellent source of foraged wood. Together, this all forms a very small agro-sylvo-pastoral system that is increasingly self-fertile.[11] After five years, we were able to see a significant amount of humus accumulating on the islands, particularly in the aisles that were permanently mulched. A layer of good compost nearly 10 centimeters (4 inches) thick had formed, and we used it to increase the volume of the beds. The soil was deep and loose, full of life. It smelled like the woods. We also had more wildlife in what had become an incredible oasis of biodiversity. Birders and naturalists rejoice at the number of rare birds living so close to constant human activity, and at the vitality of the aquatic environments.

Seeking Permaculture Farms

We began to beat our drums a bit, looking for other permaculture practioners to network with; we were convinced that there must be legions of permaculture farms in France and around the world wanting to do the same. But for years we found no opportunities to interact with truly market-scale permaculture farms. Our research in France, England, and the United States gave us the impression that most farms claiming permaculture status were essentially self-sustaining food producers but not commercial production farms. While the proliferation of home-scale permaculture farms was highly desirable, as market gardeners and vegetable growers we were interested in finding practitioners producing food in a professional setting.

Organic market-scale farms were plentiful, but extremely few incorporated permaculture practices. It seemed that permaculture and organic agriculture were two separate worlds. It became apparent that we had found our niche, to paraphrase our friend Bernard Alonso, a Canadian professor who teaches permaculture on our farm: fostering connections between organic agriculture and permaculture.

Five years later, the situation has changed. The number of projects and achievements based on permaculture in agriculture is increasing, although production farms are still rare in our country. Interest is growing, though, and more than five hundred market gardeners and project leaders have come to train at the farm. We have also been invited to teach organic market gardening.

Paradigm Shift

But back in that spring of 2009, our gradual introduction to permaculture was progressively throwing us into another world, that of microagriculture. It would take us a while to abandon our old ways, though. We were straddling two concepts of agriculture. As soon as we became farmers, we entered a world that had two widely held beliefs. The first: The bigger you are, the better off you are. The second travels in lockstep with the first: The more mechanized you are, the more you gain.

We had refused to mechanize. But we were still eager to grow — which, at our level, did not mean anything like buying a few dozen acres or the neighbor's farm. In this third year of farming, we were still earning nothing, and sincerely believed this was because we did not produce enough, because our gardens were too small. We set out to cultivate new areas. We created rows of crops between our fruit trees. And we started a more ambitious project: installing gardens on the very steep terrain that we had acquired over by the abbey.

No sane farmer would have tried to cultivate such a steep slope. To prove this point, I should share that two tractors toppled on it. The land, abandoned for fifty years, was absolutely impenetrable when we acquired it. Clearing it took two years. Then we planted an orchard with about three hundred varieties of apple, pear, cherry, plum, peach, apricot, and fig. For many years we have cultivated the less steep parts with animal power, expending great effort, but with lousy results. Working with animal traction on a slope requires relieving the weight of the tool, which tends to slide. The work was hard, and I often felt fat and old!

In the Footsteps of Sepp Holzer

We found that permaculture had much to offer on developing slopes. In fact, Austrian farmer Sepp Holzer has become renowned for his work in this area.[12] The self-taught farmer has, since 1962, transformed the Krameterhof — the family farm he took over at the age of nineteen — by following his intuition. He redesigned the mountainside farm, set on the steep slopes of a cold valley. Where neighbors plant only evergreens, Sepp cultivates a wide variety of fruits and vegetables, and grows cherry trees and vines at an altitude of 1,100 to 1,500 meters (3,500–5,000 feet) above sea level.

He achieved this by creating numerous terraces and lakes, allowing beneficial microclimates to form. Sepp breeds fish in his ponds, raises cattle and pigs, sows vegetable seeds by hand in the midst of wild plants, and does everything against the canons of industrial agriculture. Long disparaged, he has become a world-renowned expert on regenerating land devastated by industrial agriculture. He succeeded in transforming desertified areas of Spain and Portugal, implementing "water landscapes" and creating dozens of important lakes with simple and inexpensive techniques.

New Terraced Gardens

Sepp Holzer's example pushed us to create terraces on our slope. Ten terraces were formed, following the contours of the slope as much as possible. Fourteen ponds were dug, filling with rain and runoff from the paths. Compost was applied after digging the terraces, for there remained little more than loose stones. Fruit trees were planted along the embankment, alternating with berry bushes; they now provide a harvest and temper the heat of the sun in summer. Eventually, they gave their first harvest. Crop beds were created on half of the terraces; others were tilled with the horse. The land became fertile and produced more substantial harvests. Some hives went up on the edge of the woods.

After six years of effort, these gardens are starting to look great. We did not realize when starting this project that the gardens on our slope would enjoy an exceptional microclimate. Yet freezing occurs on the slope two to three weeks later than it does in the valley, and we gain even a few weeks of growing in the spring. For a gardener, starting crops as early as possible in spring is of great interest. We can begin cultivating the gardens on the hill as early as February, before descending to the valley. Aromatic plants like it there, too. And our sheep graze the steepest spaces between terraces.

These gardens, which are not connected to either city water or electricity, now function in total autonomy. For fertilizer, we take from what nature gives us — dead leaves and ferns, abundant in the surrounding woods. Nettles and wild plants on the embankments are regularly mowed and made into mulch to cover the aisles or the beds.

We no longer need fossil fuel and get what we need from nature — sun, rainwater, and the biomass that grows locally — without asking anyone for a thing.

Volunteer Reapers

Today, one thing leads to another and the work is always in flux. We now shun the use of any mechanical engine and look for alternatives. We learned to mow by hand, advantageously replacing the horrible trimmer with a scythe, handcrafted in Austria; it's the best in Europe, marketed by our friend Emmanuel Oblin.[13] The scythe is so pleasant to handle that using it becomes an addiction! The sharp blade stirs up the scent of the wild mint and oregano that thrives in spots on the slopes.

The view is beautiful, with the abbey and the village huddled below and the wooded hills on the horizon. From the towers, the bells of the abbey fill the air with powerful vibrations. Hitching up the horse, riding up to the silent workday in the gardens, and cooking lunch over a wood fire under the hundred-year-old yew tree feels like having one foot on land and one foot in heaven. This is anything but work; it's a source of inspiration, a holiday! When the cart returns to the valley in the warm light of the setting sun — full of crops, armfuls of ferns that will serve as mulch, and hazel rods to repair a fence — the intimacy of the small farm, nestled around its river, shines through. We are greeted by the animals that do not fail to acknowledge our return, each in his or her own way, and we rejoice at a new way of living.

We're fortunate to have two very different types of terrain, even though the distance between them remains an inconvenience. Both have in common the fact that they were considered unsuitable for crops. The concepts of permaculture allowed us to transform them into viable growing areas.

We have learned that permaculture design — even if it seeks primarily to enhance the land only with existing elements and avoid external inputs — often takes more work and investment than classical growing techniques for the initial transformation. Planting trees and hedges, digging ponds, and taking advantage of the landscape to promote favorable microclimates takes serious effort. But this work is also exhilarating because co-creating an edible landscape with nature is one of the greatest adventures that we could live. Once these adjustments are made, observing nature's response as she produces more vitality, biodiversity, autonomy, and luxuriance, year after year, is a constant source of wonder.

A Different Concept of Time

Finally, a permaculture approach demands a different relationship with time. Like the first peoples, we are not trying to maximize short-term profit, but rather to seek balance over time. If I die tomorrow, I will have worked hard and reaped little in the gardens on the hill. But those who come after me will live for generations in a beautiful place, work a healthy land and self-fertilized gardens. If the world enters troubled times, they can feed their hunger. I have had the joy to create these gardens. This is already a lot.

NINE

Biointensive
Microagriculture

The human body is still more efficient than any machine we have been able to invent ... Using hand tools may seem to be more work, but the yields more than compensate.

— *John Jeavons*[1]

We had our first experience with raised beds during the 2009 and 2010 seasons. In the end, we were happy with our results, but we must admit that we came across several significant obstacles in search of reliable data. The first was the absence of professional gardening resources that covered raised-bed cultivation. We read a lot of books on permaculture, but the literature out there was primarily geared toward people who want to develop their edible landscape on a home scale.

We found little that answered our questions about the subtleties of raised-bed cultivation. Was it true that you never really needed to work the soil in a bed? Would keeping it covered with a mulch, and letting fertility increase naturally, really be enough? After two seasons, we found that some of our beds were compacted, others not. Those that had the benefit of a significant contribution of compost remained flexible and airy. Now we decompact our beds in spring with a broadfork, and repeat the operation if necessary between plantings.[2]

We also had questions about fertility. We often found contradictory statements in professional organic gardening literature, permaculture books, and testimonies from old-timers. "No need to add compost," declared a permaculture instructor; "decomposed mulch is enough." I confess to being uncomfortable with black-and-white statements — then and now. Maybe he was right, but maybe not. There were so many parameters to consider. The fundamental question of fertility seemed to deserve more nuanced answers. What is the initial fertility of the bed? How many different crops does it produce?

Paradise, for Slugs

As we sought answers, our gardens became prone to a population explosion of slugs and voles, for whom mulch is a godsend. We had created a garden of Eden — for slugs. The damage they caused was significant, making salad greens and sprouts unfit for market. We had to find solutions to deal with these downsides of mulch. It turned out that, like most of the strategies we explored, mulch management is based on a set of varied and complementary measures.

Natural farming was taking us into a world of increasing complexity — not surprising, since life itself is complex. We had to let go of standards, rules, and prescriptions, and enter into a detailed observation of each crop and its interaction with its environment. If there are rules, we now realize, they should be metarules, conceptual benchmarks to creating a bio-inspired system. This perspective is fairly new in the world of contemporary agriculture, since one of the characteristics of agricultural production is that the farmer has, increasingly, become a performer of directives from technicians.

But each locale is unique. Each farmer, too. And permaculture attaches great importance to the fact that a project should be designed to best match the place and the people who are doing it. The author who inspired us the most during these first two years was Patrick Whitefield, a farmer and English permaculturist. One of his rich books is *The Earth Care Manual*, a work of 470 pages with dense, clear text, based on a permaculture approach adapted to temperate European countries.[3] It addresses interesting nuances for a professional gardener growing in colder European countries.

Accept Being Small

Another thing that made our situation uncomfortable then was the fact that the working ratio between our cultivated area and what we could actually maintain was not good. The surfaces of our beds were too large to be cultivated in a very careful manner: There weren't enough of us to keep up with the work. To get the best out of your beds, the care factor is absolutely crucial. We had continued to expand our gardens, thinking we were still too small to earn a decent living from our business. We had not yet realized the potential productivity of perrenial raised beds. Working by hand, it is impossible to properly treat a large area. A small, very well-cultivated area will be more productive than a large one that is not maintained. But this we had not yet understood.

There is an inherent logic in each approach to agriculture, which must be understood. Otherwise, a farmer will experience the disadvantages of each system. The mechanized organic market gardener may, with a tractor, quickly prepare an acre, cover it with plastic using a sheeting reel, and transplant thousands of plants with a mechanized planter. The same amount of time spent on raised beds only allows us to prepare a few dozen square-meter plots. If we seek to do more, as with the tractor, we are doomed automatically. It is pointless trying to match a machine more powerful than us. It makes more sense to explore what we can do with our hands that a tractor cannot accomplish.

Poorly maintained crops will be covered with weeds and subject to erosion, and all the hard work of bed construction, fertilization, seeding, and transplanting will have been in vain. You will harvest nothing and almost everything will have to be redone. A garden bed is a space that is costly in labor and justified only if it has a bountiful harvest.

Eventually, we would start to find the right balance, moving closer to the most productive surface size and configuration — something that we'll come back to in the pages ahead. We cannot overemphasize how important it is to achieve peak productivity. It can determine the success or failure of a farm.

John Jeavons, the Man Who Grows More Vegetables

After two years of growing with raised beds, we looked for new influences. Exploring alternative agricultural practices became, once again, an exciting journey through countries, cultures, and time. We did not imagine embarking on such an involved investigation to become farmers!

A milestone in this journey was reading John Jeavons's book *How to Grow More Vegetables than You Ever Thought Possible on Less Land than You Could Imagine.*[4] I must confess that reading this title made me skeptical at first. Also, with the cover bearing the words "more than 500,000 copies sold," I thought at first that it seemed too commercial, too American, and I put it aside for a year before opening it. What a pity! I could have been reading the result of forty years of research by his nonprofit organization, Ecology Action. Its pages are most instructive.

A Bit of History

In the 1920s, a young and talented English gardener, Alan Chadwick, delved into the heritage of horticulture on the old continent to unearth its traditions. He trained with former Parisian market gardeners who had developed surprisingly productive techniques in Paris and the suburbs. Chadwick also studied biodynamics from its founder, Rudolf Steiner. Steiner formulated the foundations of biodynamic agriculture in a series of eight lectures called the Agriculture Course, in 1924. Chadwick synthesized the two approaches and called it "the biodynamic French intensive system." He perfected his practice for some fifty years, in Europe, Africa, and America.

At the beginning of the 1960s, Chadwick created a garden with students from the University of California at Santa Cruz. Now called the Alan Chadwick Garden, it is maintained by the university's Center for Agroecology and Sustainable Food Systems. (Perrine was able to visit this exceptional school.) In three years, they transformed an arid and uncultivated land into a thriving little paradise, where Chadwick began teaching his method. This great horticulturist had the honor of passing a rich tradition from the European continent to the United States. Curiously, while the heritage of former Parisian market gardeners had been forgotten in France, their knowledge was being further developed in the United States. In the midst of the country's agricultural gigantism, Chadwick affirmed proudly: "Just grow one small area, and do it well. Then, once you have it right, grow more!"[5]

In 1972, a young team formed by Chadwick created a microagriculture research and teaching center managed by Ecology Action. A pillar of the organization, Jeavons continues to study, practice, publish, and train others well into his seventies. Ironically, his method has remained little known in France until recent years, and only one French person, Rachid Boutihane, was trained by him.

A Damning Report

John Jeavons and his team named their growing method GROW BIOINTEN-SIVE. He describes the approach as agricultural miniaturization, an attempt to counter the problems brought on by industrial agriculture — which, he stresses, destroys the soil at an accelerated rate, losing between 6 and 16 tons of soil for every ton of food produced. Mechanized organic farming, according to Jeavons, is little better: It destroys the soil seventeen to seventy times faster than nature creates it. By buying food that is cultivated at the expense of destroying topsoil, writes Jeavons, we become complicit in this destruction. He stresses that, according to various studies, if we continue to destroy soil at this pace humanity will have degraded all the arable land on the planet in the next century. The biointensive method creates soil at a substantial rate.[6]

The arable land that each person has to meet his or her nourishment needs is shrinking. "It is becoming increasingly clear," notes the Ecology Action team in the preface to Jeavons's book, "that GROW BIOINTENSIVE Sustainable Mini-Farming will be an important part of the solution to starvation and malnutrition, dwindling energy supplies, unemployment, and exhaustion and loss of arable land . . ."[7]

A Key Question

Therefore, Jeavons asks this crucial question: "What is the minimum surface on which a person could grow enough crops to give him all of his food, clothes, building materials, compost, seeds, and income for a year?"

It is difficult to answer this question, as the data may vary from one place and from one person to another. But this is the kind of questions that soon we no longer will be able to avoid, when the scarcity of resources will force us out of globalization and require us to meet our basic needs locally. As for our food needs, the forty years of research by Jeavons and his team suggest that with biointensive microagriculture, "about 370 square meters cultivated is enough to grow all the food needed by a person for a year" while providing all the materials for compost needed to maintain fertility. This calculation is based on a vegetarian diet. Why then look for macro responses to the enormous challenges of our times? asks Jeavons. Rather, try to develop micro-responses at the individual level, to meet our own needs.

The research of Ecology Action is focused on individual self sufficiency rather than on the field of professional agriculture. Yet the potential of microagriculture is such that Jeavons's approach has inspired diverse development programs throughout the world, making farmers more likely to cultivate small surfaces. In the 1970s, Jeavons believed that areas of 500 to 2,000 square meters allowed net incomes of $5,000 to $20,000 per year. A woman from British Columbia earned about $400 a week by growing vegetables for restaurants on only 225 square meters.

These numbers would surprise most French market gardeners. We are not used to assessing the productivity potential of microagriculture. The study that we have undertaken at La Ferme du Bec Hellouin will confirm these results in our own context — the France of the 2010s.

Building the Soil, Preparing Our Future

Jeavons quotes Gandhi: "To forget how to dig the earth and tend the soil is to forget ourselves."[8] The attention given to maintaining and developing cultivated land is one of the most interesting points of the biointensive approach. For Jeavons, it is important that each garden creates the conditions for its own fertility. A thorough study, taking into account both the nutritional needs of people and the nutritional needs of the soil, led the Ecology Action team to define a simple rule: Part of the terrain must be devoted to high-yield crops that produce biomass used to make compost for fertilizing the entire garden. Thus driven, the garden can be sustainably self-fertile.

The golden rule of biointensive microagriculture is to divide cultures as follows:

- Sixty percent of the cultivated area is dedicated to biomass plants, able to provide the bulk of the materials to be composted for the whole garden. These plants — such as grains, beans, and sunflowers — procure carbon (the main component of the organic material) and calories.

- Thirty percent of the cultivated area is dedicated to tubers and other vegetables rich in calories, such as potatoes, leeks, Jerusalem artichokes, garlic, parsnips, sweet potatoes, and oyster plants.
- Ten percent of the surface is dedicated to various vegetables that provide vitamins and minerals: salad greens, carrots, radishes, and turnips, for instance.

It should be noted that grains are grown in raised beds and usually transplanted. In Europe, wheat was once commonly grown in gardens with yields of up to 100 quintals per hectare (against about 70 today; 1 quintal is about 200 pounds), without any form of mechanization.[9]

An Eight-Point Approach

The GROW BIOINTENSIVE method was formulated in eight points.[10]

1. **Deep soil preparation, double digging.** Double digging is an old gardening technique that unpacks the ground two shovel-scoops deep. In the biointensive method, the soil layers are not mixed; this technique allows the formation of permanent garden beds with ideal conditions, including fertility and deep decompacting.
2. **Composting.** Self-fertilization is the goal. Composting is accomplished by layer, carefully, alternating beds of carbon-rich material (branches, leaves, straw), nitrogen-rich material (grass, peelings, and food scraps), and earth.
3. **Intensive planting.** Vegetables are planted systematically, staggered to fit more crops and create a favorable microclimate between the soil surface and leaves.
4. **Companion planting.** Crop associations foster plant diversity that keeps the garden healthy while raising its level of productivity.
5. **Carbon farming.** Plants absorb carbon from the atmosphere as they grow. Once they are composted, that carbon is stored in the soil.

6. **Calorie farming.** Producing calorie-rich crops allows farmers to maximize their efforts toward providing for their food needs from small areas.

7. **Open-pollinated seeds.** The gardener looks for self-seeding varieties (pollinated by insects, birds, or wind) and favors older varieties.

8. **A whole-system gardening approach.** The method must be performed in its entirety, as a matter of consistency and efficiency.

Effectiveness of Biointensive Microagriculture

The GROW BIOINTENSIVE method was evaluated for decades. According to Jeavons, the results speak for themselves. He cites in his book the following data:

- The yields obtained by the GROW BIOINTENSIVE method are, on average, two to six times higher than those of US agriculture, and up to thirty-one times higher for some crops.
- Water requirements are reduced by 67 to 88 percent per unit of production.
- Purchased fertilizers (that is to say, not produced on the farm) are diminished by at least 50 percent — or even obliterated.
- Fossil fuel use is reduced by 94 to 99 percent.
- The calories generated per unit of area are two hundred to four hundred times higher.
- Income per unit of area is at least doubled.

It is clear that a manual approach to agriculture may well shake up preconceived ideas. Many may think that mechanization increases efficiency, but does it really, if crops are obtained at the cost of huge losses of topsoil? John Jeavons asks the real questions and proposes viable alternatives to the dominant system, accessible to everyone.

Applications at La Ferme du Bec Hellouin

Careful and regular reading of Jeavons's book has helped us gain confidence in the potential of microagriculture. We cannot say this enough: The obstacles have been mainly in our own heads. We've had to decondition our mental formatting to evolve our practices.

Sufficient soil thickness was a problem on our farm. Only in the vegetable garden in front of the house were we able to test the double-digging technique. Constructing a permanent raised bed with this technique is obviously a plus, since plants benefit from decompacting the soil between 60 and 80 centimeters (24–32 inches) deep, deeper than the most powerful tractors can go. Working time to build the bed is approximately tripled. We have not observed a difference in returns relative to our other beds, but other factors are likely to influence the results.

Jeavons and Chadwick emphasize meticulous care for crops at every step: preparing the soil, producing seedlings, transplanting, and so on. This care factor is one of the parameters that creates a "competitive advantage" between a market gardener and a tractor. Microagriculture is only productive if the work is very neat. This goes against a certain train of thought that is in vogue in permaculture circles, and argues that natural farming is to "let" nature take its course, and that there is "little work" for the gardener. I must admit that I am doubtful. That may be true for someone who just wants to cultivate something to feed themselves, but to produce sixty vegetable baskets every week, throughout the year, I'm skeptical. That taking advantage of the ecosystem allows, eventually, for a reduced workload is undeniable. Going with the flow of life, trying to understand what is good for the soil, good for vegetables, good for people and for the whole world of the living, is the foundation of natural farming. But nature does not spontaneously grow the sophisticated plants that are our vegetables. The plants that we are used to eating are usually the result of a long co-evolution between a wild plant and humans; they need our care and demand fertility, water, and sun.

Perennial Plants, Wild Plants

Leaving more space for perennial plants, which consume less labor and fewer inputs, makes a lot of sense. In nature, a vast majority of the plants are perennials; a Western diet made up almost exclusively of annual plants is nonsense — like any specialization, it's risky. So we aim to expand the range of perennial vegetables grown on the farm. The spinach-like patience dock

(*Rumex patientia*), for example, is a hardy perennial that grows very fast in early spring and allows several successive cuts.

It stands to reason that wild plants should be part of our diets. Besides the fact that they grow by themselves without asking for anything from anyone, they are generally more concentrated in active ingredients and nutrients than our selected vegetables.[11] But the market is currently very limited. Professional growers are required to meet the expectations of their customers and cannot hope to revolutionize their eating habits overnight — though they can, perhaps, foster change over time. Meanwhile, our market garden offerings must be beautiful and must also include the conventional fruits and vegetables that have become integral components of our short-sighted food culture.

Permaculture has everything to gain by enlisting the approaches of Jeavons, and vice versa. In my humble opinion, the GROW BIOINTENSIVE method can be enriched by permaculture concepts. Some important points — the overall design of the garden, microclimates, the role of trees and water, the benefits of a permanent mulch, and more — are touched on either briefly or not at all by Jeavons. Perrine and I tried to incorporate the wonderful techniques of biointensive microagriculture in the broader context that permaculture offers, and the results have proven very satisfactory.

There is, however, another great North American pioneer who inspired us to an even greater extent: Eliot Coleman.

TEN

Eliot Coleman

I am passionate about climbing. As a climber, I can say what is incredible about agriculture is that it is like a mountain, except it has no top.

— *Eliot Coleman*[1]

$\mathscr{N}$o doubt about it: For us, the most significant book of recent years was *The Winter Harvest Handbook* by Eliot Coleman.[2] Perrine bought it in England. A masterpiece! This book caused a mini earthquake that had a profound impact on our farm. As soon as we received it, I spent hours gazing at pictures of greenhouses and gardens, blown away by the beauty that emerges from Coleman's Four Season Farm, where the layout of crops touches perfection. I devoured the text, clear and full of useful information. Over the years the book became wrinkled, stained with earth, covered with annotations and calculations to convert to the metric system from feet and inches. Coleman is one of the best growers in the world. Perrine and I have for him the same veneration as a *sadak* Hindu has for his guru.

After hearing about his methods in our training, one of our interns, Elsa Petit, offered to translate the text to French. Encouraged by our enthusiasm, she worked briskly. We then showed it to Françoise Nyssen and Jean-Paul Capitani, directors of the French publisher Actes Sud. The two are passionate about everything related to agroecology (Jean-Paul studied agronomic engineering) and they agreed to publish it in France.

Then, in autumn 2013, for the release of the French edition, Eliot visited our farm with his wife, Barbara Damrosch, also a talented gardener, to give a dazzling training session to seventy people. These wonderful days were also spent in the company of Philippe Desbrosses, a pioneering organic farmer in France who is also a guiding spirit. Philippe and Eliot had met thirty years earlier in the international organic agriculture community. These two youthful septegenarians — and Barbara, too — showed us what fresh air, enthusiasm, and commitment in life can accomplish.

The Market Farmer Life

It's always a privilege to meet people who live their dream. From our first meal together, Eliot and Barbara were thoroughly interrogated. "When I was a kid," said Eliot in almost perfect French, "I had so much energy . . . These days they would sedate a kid like me!" His first profession was teaching, but he seemed to have been more interested in the outdoor life than classrooms. A sports enthusiast, he spent his time climbing and canoeing, always in search of challenges. Pushing limits was his obsession. Simply being told that a mountain was impossible to climb would have him packing for the ascent. "We climbers are always seeking the elegance of the way." It is with this attitude that he was able to masterfully develop his farming approach: accomplishing much with great economy of means and without apparent effort.

In the late 1960s, Eliot was thirty years old. After being a conscientious objector and wandering from one mountain to another, he was faced with a major challenge: to transform an evergreen forest, acquired for the "handshake" cost of $33 an acre from Helen and Scott Nearing, into a farm. The Nearings were simple-living pioneers whose work focused on rekindling self-sufficiency skills, and they wanted to help Eliot start his farming adventure. Eliot began chopping down trees with an ax to open a clearing in the forest, discovering a floor of sand and pebbles at pH 4.5 — very acidic and unsuitable for vegetable crops. There was nothing particularly original in this: The history of the United States is full of crazy pioneers implementing the recommendations of President Abraham Lincoln: "Population [will] must increase rapidly, more rapidly than in former times, and ere long the most valuable of all arts will be the art of deriving a comfortable subsistence from the smallest area of soil."[3]

But where Eliot distinguished himself from the pioneers of yesteryear was in the fact that, twice monthly, he made a three-hour drive to the library of a university and devoured everything that related to natural agriculture.

Soon his unquenchable curiosity pushed him to fly to Europe. In France, Eliot set out to meet the remaining heirs to the rich tradition of Parisian market gardening. In the fall of 1974, he was shocked to discover, in Blain-villiers, in the southern suburbs of Paris, the market gardens of Louis Savier. "That visit to Savier's garden was the most powerful influence on my development as a market grower," writes Coleman. "My handwritten notes begin with one word — 'wow!' Quality was everywhere: the organized layout, the tidy closely spaced rows, the ranks of cold frames and hotbeds, the dark chocolate-colored soil and, most especially, the crops glowing with health. Even though I had begun my own market-garden career six years earlier and had read many books about intensive production, it wasn't until I stood in Savier's garden that I realized how well it could be done. I visited Louis Savier on three subsequent occasions prior to his retirement in 1996. Each visit was more impressive than the one before."[4] France became an ideal stage for Eliot's increasingly frequent study tours.

His career then took a few turns. He directed an experimental farm school, became the director of IFOAM (International Federation of Organic Agriculture Movements) for two years, advised on various projects, and designed innovative tools before returning to his Four Season Farm in the late 1990s. Eliot had already authored two popular organic agriculture reference books, *The New Organic Grower* and *The Four-Season Harvest*, when he began writing *The Winter Harvest Handbook*, a ten-year project.[5] He had also gained much experience in international microagriculture methods and applied them on Four Season Farm, which had become a model microfarm — beautiful, productive, and generous. Under its tutelage, tens of thousands of small farms have sprouted up across the United States, where the organic microfarm movement is now firmly established. In France, the movement remains in its infancy.

The Four Season Farm

The farm is located in the northeastern United States, on the Maine coastline, a harsh place where the thermometer regularly drops to -13 degrees Fahrenheit (-25 degrees Celsius) in winter. Eliot operates it year-round, but specializes in growing excellent vegetables in the heart of winter — infinitely more difficult and more technical than growing summer crops. He was one of the first to develop the techniques of using double coverings on crops and movable greenhouses.

Eliot's primary concern is his soil. Though he started out with soil that was unsuitable for crops, he has patiently transformed it into wonderful market farm soil. "When I do soil testing," he quips, "it is like a report card: I have A+ everywhere!" It took him many years to get this far, incorporating green manure with significant and regular additions of compost. The farm today is cultivated so intensely that there is no more room for green manure in the rotation, but Eliot always takes great care with his compost.

He shared a very interesting observation with us: When soil becomes healthy, alive, and well balanced, pests disappear. He has on many occasions found that a disturbance in the soil has led to the return of pests. He has on occasion removed a big rock from the middle of a bed and filled in the hole with surrounding soil. The next crop is prone to infestation in this spot. This observation really struck me. It's yet another reason to focus on caring for the soil, which is the live foundation on which the life of a small farmer is built.

Eliot has actually *created* his soil, reviving the practices of the old Parisian market gardeners. In modern literature covering commercial-scale organic agriculture, we no longer talk of creating the soil. How could we, since we now work mechanized equipment over large areas? One hectare (2.5 acres) generally has 2,000 to 7,000 tons of soil, which would be extremely difficult and onerous to transform. This is one more argument in favor of microagriculture: Cultivating small areas as permanent beds makes it possible to transform a barren substrate into excellent-quality land. We can't emphasize enough that one of the secrets of productivity for microfarms is attention to the soil. One becomes a creator of humus.

The Secrets of High Productivity

One of the most interesting aspects of Coleman's work is the great logic and unity of a system based on permanent 75-centimeter-wide (30-inch-wide) beds. By choosing to avoid mechanization as much as possible, Coleman was able to pack his vegetables more closely together.[6] The tractor necessitates planting in long rows, and also dictates the width of the pathways between rows. (They must be wide enough for the tractor wheels and tools to pass through.) Eliot searched for ways to plant the crops as densely as possible. Progressively, he diminished his pathways and now leaves intervals of just 6.5 centimeters (2.5 inches) between small vegetables like carrots, radishes, and young shoots. He designed a wonderful manual six-row planter that, in two passes, can sow twelve rows of vegetables with a high degree of accuracy. As plants grow, their leaves quickly spread out to touch one another, covering the entire surface of the bed, preventing erosion, and limiting weed growth.

We can see the advantage of this precision seeder by comparing it with a tractor, which can plant only about three rows at a time in a 75-centimeter bed and leaves the land between the rows bare, leached by rain and burned by the sun, an open door to the development of weeds.

Do Not Invite Pioneer Plants

Weeding is one of the great challenges of organic farming. It's worth taking a deeper look into what we commonly call "weeds." Weeds are generally wild plants that tend to thrive on degraded soils. Recall that, in nature, soil is never naked unless some sort of mishap has made it so. These wild, pioneer plants therefore play an important role: They are able to grow on very poor soil, withstand drought and heat from the sun, and regenerate soil, which allows it to later accommodate more delicate vegetation, like our crops. They are programmed to grow very quickly and withstand hostile conditions. Every time we leave bare spots in our gardens, we send a call to pioneer plants to quickly colonize to repair them. When we better understand the function of these plants in nature, it becomes easier to avoid favorable conditions in areas where we want to keep them at bay — or conversely to regard them as allies and benefit from their vitality.

In other words, as Eliot often stresses, don't invite plants you don't want by leaving places for them to grow. Which brings us back to his manual six-row planter. Not only does it minimize the spaces where unintended plants can spring up, and sow many more rows at a time, but this simple tool makes powerful economic sense, too. Its initial cost is less than $600, compared with the extraordinary cost of a tractor, and its operating cost is zero. Coleman writes that when he switched from a conventional model to the precision six-row seeder, it nearly doubled his yields. I challenge anyone to find another agricultural innovation that allows double returns for $600.

But the value of this seeder is equally high on the ecological front. The system of permanent flat beds for which it is designed respects the soil, allowing it to be densely inhabited by roots and microorganisms that will naturally make it more vibrant.

This tool is still perfectable. Eliot is working with Johnny's Selected Seeds, the tool and seed company that produces and distributes his creations.[7]

A Cohesive and Efficient Agricultural System

For the tractor, Eliot has gradually substituted a variety of hand and power tools specifically designed for his system: narrow hoes to work between tight rows; a mini tiller, powered by a battery-operated electric drill, that works only the first 5 centimeters (2 inches) of soil and does not unearth the buried weed seeds; and a harvester for young shoots, also run by a battery-operated drill. And that's likely not all: Eliot scribbled out for us his plans for a versatile electric tool for performing different tasks on permanent beds. He still has plenty of ideas.

The yields of the Four Season Farm are such that its cultivated 8,000 square meters provide wages for two permanent and four seasonal workers in the summer, in addition to Eliot and Barbara. Eliot is satisfied when the turnover in the cultivated area reaches $25 a year per square meter. The revenue per hectare (2.5 acres) averages $200,000 per season. In France the average (non-biointensive) organic market farm averages just 30,000 euros (roughly $33,750 at the time of this writing) per hectare per season. That comparison should convince those who still doubt the effectiveness of the biointensive market garden. It also reflects the fact that organic vegetables bring significantly higher prices in the United States.

Behind the Techniques, an Ethic

If Eliot has designed and developed a high-performance technical system, it's because he is animated by a vision. He wants to inspire and instruct people to create more microfarms that feed local communities with healthy and natural products, while providing a good quality of life for farmers. The nonviolent commitment of his youth transformed into an agricultural practice that's respectful of the land and people. "Capitalism and communism tried to destroy small farms because they are independent," he believes. "But I see them as the future of agriculture."

Eliot still runs his farm and has no intention of stopping. "No retirement," he declares, "I will die in my garden! If I stopped working the earth, I would also cease to write. I only want to write about what I practice." Barbara is equally invested. She wakes up at 4 AM to work in the garden and write her books or her weekly *Washington Post* column, and finds time each Friday to cook for the whole Four Season Farm team.

Eliot is always looking toward the future and gives all he can of his time and ideas. None of the tools he invented are patented; he wants them to be copied and serve the small-organic-farm movement that he holds so close to his heart. "Copy them! Improve them!" he advises us. Generosity is one of the characteristics of this good man who came to us loaded like Santa Claus with beautifully designed tools. He has sold plots of land to his workers for barely anything, in the spirit of the low price he paid forty years ago.

During the three days we spent together, we could see that Eliot is a market gardener through and through — to the end of his fingernails, as we say in French. On a tour of our gardens, he frequently leaned in and touched the soil, closely observed the vegetables, and advised us about the different varieties. I loved watching his eyes narrow and his focus intensify when a subject interested him. We were surprised and touched to discover many common interests and spent long evenings talking, until Philippe Desbrosses would take out his guitar to fill the air with song. Eliot is not the type to preach but, when leaving us, he talked about his relationship with nature, his pacifism. "In the mountains, I learned not to be afraid of death. The fear of nature is linked to our fear of death. I look at it differently. I'm not afraid of insect pests, for example!"

As he started his car to leave La Ferme du Bec Hellouin, he said jokingly, "Your farm is magnificent! It justified the trip; you are a power couple! I'm going to plant trees in my garden. Try not to cut yours before my return."

Applications at La Ferme du Bec Hellouin

We discovered Coleman's work at about the same time as that of Jeavons, after two years of cultivation in round garden beds. We became aware of some shortcomings of round beds, particularly for direct seeding. It was impossible to use our mechanical seeder (in a row) in the round beds. Seeding by hand, "on the fly" is more random: You can get a little heavy-handed, or sometimes you sow too lightly, which does not promote high and steady returns.

When we discovered flat beds and the six-row planter, we were quite game to adopt it, but didn't want to entirely give up the round beds, which we greatly appreciate. Since then we have been developing a system of round beds of various sizes and standard flat beds 80 centimeters (32 inches) wide. This suits us perfectly; each type of bed has advantages and disadvantages.

The flat beds, which we call Coleman beds, are ideal for direct seeding. We also use them for transplanting. They require more care than the round beds and are less self-fertile, usually requiring the addition of compost between each crop. All our greenhouses have flat beds as well.

The round beds are more "natural" and require less intervention. Their shape makes them perfectly suited to transplanting, but sometimes we continue to do direct-broadcast seeding. We try to keep the beds covered with mulch as much as possible and no longer add compost.

The two systems are complementary; one is just more anthropomorphized than the other.

However, we have allowed ourselves to modify the system of our guru, Eliot. Influenced by the permaculture approach, we lay mulch on flat beds, when possible, for crops that take longer to mature — like tomatoes, cucumbers, peppers, and eggplants.

Our main departure from his system, though, is that we plant several crops in the same bed. I have often wondered why Coleman does not take advantage of companion planting, which was used nearly systematically in the Parisian market gardens of the nineteenth century that inspired him. When I asked him this question, he said that his rotations are very studied and complex and he does not want to complicate them further.

It seems to me that the economic results we get are largely the result of these crop associations. We sow, for example, twelve rows of carrots in a bed, followed by twelve rows of radishes. Then we transplant a row of salad in the middle. The radishes are harvested first, then the lettuce, which is replaced by cabbage. We merely reproduce a classic combination of the nineteenth century. And it works!

In the greenhouse, we plant lettuce between the rows of summer crops (tomatoes, cucumbers, peppers, eggplants) and sow radishes and turnips along the border of the beds. Then the lettuce is replaced with basil shoots. In August, when summer crops are in full production and basil has run its course, we replace it with (for example) celery, which will be harvested in the fall two weeks after the end of the summer crops. Thus, instead of having a single crop, we manage four or five over the same time period in the same bed. Their little worlds coexist without bothering one another too much.

Everyone develops a system that suits their style. Eliot Coleman is remarkably organized and meticulous; his grounds are impeccably ordered. Personally, I like a system that is a bit more in flux, where cultures mix, with trees everywhere; otherwise I get bored. Perrine prefers something wilder yet: She respects weeds and hates it when I cut the nettles in her forest garden. We enjoy carrying out a good number of original experiments, even if they do not work as well as we would like sometimes. It allows us to stumble across a gem once in a while.

Everyone should feel comfortable and fulfilled by the system they use. It is worthwhile to forage here and there through "best practices" and concoct your own synthesis. This was brilliantly realized by our friend from Quebec, Jean Martin Fortier.

The Market Gardener from Quebec

Perrine, during one of her evenings on the Internet, found a small farm in Quebec, Les Jardins de la Grelinette, which uses essentially the same techniques and tools as we do — those of Coleman — and gets some very respectable results. We began to correspond with its kind creator, Jean-Martin Fortier.

In the fall of 2012, Jean-Martin published a practical manual of great interest to us, *The Market Gardener,*[8] describing in detail the operation of his farm. He is an excellent market gardener, and equally at ease in his role as teacher. With his partner, Helena Maude Desroches, he managed to establish a microfarm with surprising economic results in the agricultural world of Quebec. In the spring of 2013, we were happy to welcome Jean-Martin to La Ferme du Bec Hellouin for training and good times, sharing thoughts on our farming practices. Jean-Martin describes himself as a family-focused market gardener. He has a more relaxed attitude toward making money than many Europeans, and he shares freely that their main interest in creating their farm was to set up a profitable operation that allowed them to take several months off in winter. The relatively important fertilizer purchases, the heating of greenhouses in the spring, and the use of small mechanical appliances all fit into this vision. Maude and he quickly reached their goal: The production is abundant, with beautiful vegetables and loyal customers. They created three jobs on less than 1.2 hectares (3 acres), and ultimately the ecological impact of their farm is much smaller than that of mechanized vegetable farms of equal size.

We strongly recommend the books by Coleman and Fortier to all those interested directly or indirectly in microagriculture. They will set you ahead years in your practice. (We hope that French literature, lagging a generation behind on these issues compared with the Anglo-Saxon world, will soon come up to speed as well.) Without having planned for it, La Ferme du Bec Hellouin has served as a link between the two shores of the Atlantic, and we thank our American friends for sharing with us so many discoveries. One of the most exciting things we discovered through them was our own heritage, that of the Parisian master gardeners of the nineteenth century, covered in books by both Jeavons and Coleman.

The Hervé-Gruyers constructed this permaculture school on their farm using old materials, but it is a bioclimatic building completely powered by renewable energy.

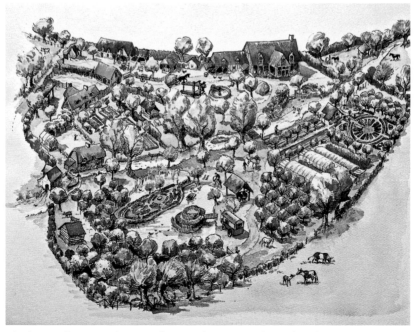

The design of La Ferme du Bec Hellouin took years to evolve and is still a work-in-progress.

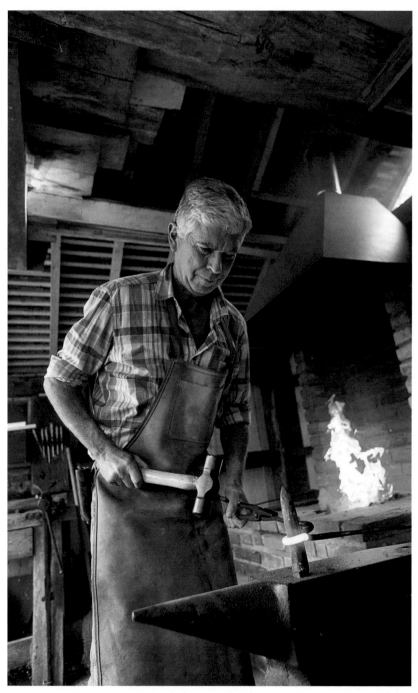

Charles at work in the forge. *Photograph by Claudius Thiriet.*

Minimizing the distance between rows, associating crops, and using vertical space accommodates more plants on less acreage.

Children get much satisfaction doing real chores on the farm.

Perrine processing the abundant fruit harvest — in this case apples to make cider. Hundreds of trees have been planted at La Ferme du Bec Hellouin.

Horse power replaces fossil fuels for heavy hauling on the farm.

Soil preparation is key. This simple but specialized tool created at the farm allows the beds to be prepared quickly by hand.

When chefs, including some of Paris's finest, started coming to the farm to seek out its distinctly flavorful crops, we realized our attention to detail was having greater than expected results.

What began as a quest to grace the family table with fresh, healthy food grew into a full-scale experiment in finding the most ecological and productive way to farm.

Perrine in the farm's onsite store.

Research at the farm tracks how much produce a 1000-square-foot plot can yield.
Photograph by Carlo Bettinelli/Fruzsina Gyertyan.

Animals on the farm provide constant pleasure for the children while they learn how to care for the animals.

ELEVEN

The Parisian Market Gardeners of the Nineteenth Century

Produce a lot on a small space, provide a vegetable diet for 1,000 individuals by cultivating an area that once did not even nourish 50 when ordinary farming procedures were applied to it, and if art did not come to the aid of nature — such is the problem posed every day to the culture of the market gardener in the surrounding area of Paris, and the problem every day has its solutions.

— I. Ponce[1]

The intensive cultivation of vegetables, as practiced in professional gardens where water and compost are readily available, differs from usual vegetable growing in the sense that it must be an ongoing process throughout the year, often with many different vegetables planted together on the same piece of land.

— J. Cure[2]

*W*e were surprised, during our investigation, to find that we had nothing to invent, that many interesting agricultural techniques were already in practice, sometimes for a thousand years or more. They simply needed to be rediscovered, adapted to our environment and our time, enriched by the sciences of contemporary living. The magic of modern times allows us, no matter where we are or what level of education we've attained, access to unlimited amounts of information. The main challenge is not to drown in complexity. At La Ferme du Bec Hellouin, we strive for a cohesive synthesis. We try to identify strong traits from various agricultural systems that flow with the natural direction of life, and weave them a general set of rules capable of inspiring new achievements, and of adapting to the soils, climate, and human and social context of our farm, as well as others.

"Biologically intensive farming dates back to four thousand years ago in China, two thousand years in Greece, and a thousand years ago in Latin America," writes John Jeavons. "In fact, the Mayan culture grew food this way at their homes on a neighborhood basis. This is one of the reasons their culture survived when others around them were collapsing."[3] Why were certain forms of agriculture able to feed a large population for centuries without exhausting the soil while other models converted farmland to desert?

Among the many forms that microagriculture has taken in the world, one of the most successful was practiced in Paris and its suburbs in the nineteenth century, until the rise of urbanization and the arrival of mechanization sank this rich tradition into relative obscurity. Paradoxically, the Anglo-Saxon world has maintained a great admiration for these intensive market farms of Paris, probably because of the London market gardeners who traveled to study the practices of their Parisian colleagues, and several English-speaking authors, including William Robinson and Prince Pierre Kropotkin, who wrote about them.[4] Do those who ride line 9 on the Paris Metro know that Maraîchers station, opened in 1933, was named in honor of the many market gardeners who grew fruits and vegetables on the nearby hills of Belleville and Montreuil?

Parisian Market Gardeners: Important Precursors

Parisian market gardeners of the nineteenth century reached a level of excellence that few contemporary gardeners attain. Who, nowadays, can boast of achieving up to eight or nine crop rotations per year, and produce, without any form of mechanization and without a drop of fossil fuel, salad greens all year, along with melons, cucumbers, strawberries, and tomatoes as early as April and May?

At a time when many elected officials dream of reintroducing agriculture to the heart of urban and suburban areas — it is even a global trend — the

Parisian gardeners offer an extraordinary precedent. Throughout the second half of the nineteenth century, Paris was self-sufficient, with all of its vegetables produced inside the city limits. Parisian produce, diversified and high quality, was even exported to London markets.

Little is known about the history of Parisian market gardeners before the French Revolution. The art of fruit and vegetable cultivation had leapt forward in the time of Louis XIV, under the leadership of master gardeners like Jean-Baptiste de La Quintinie, who experimented in the kitchen garden of the king, where he began to grow vegetables under season-extending glass frames and bell-shaped cloches in the 1670s and 1680s.[5] A new professional category, market gardeners, then gradually developed in the heart of Paris. They formed a "caste" whose knowledge was passed from generation to generation. In 1780, a market gardener by the name of Fournier was the first to use a frame for his crops, thus opening a new era for the profession, that of forced cultivation, conducted under cover, with a source of natural warmth.

From Firsthand Accounts

The first detailed testimony about the life and practices of market gardeners was the *Manuel pratique de culture maraîchère*, a moving document written by two practitioners, Moreau and Daverne, and published in 1845. This book exhaustively describes the crops, tools, and even the organization of the profession at the time. Here is the description that Moreau and Daverne give of their colleagues in the capital, which was in turmoil throughout the nineteenth century: "The market gardeners of Paris form the category of workers which is the hardest working, most consistent, most peaceful of all those who live in the capital. However solid or weak their situation, we never see the gardener change occupations. The sons of a gardener become accustomed to work, under the eyes and with the example of their father, and almost all become established market gardeners. The daughters rarely marry a man from a different profession than their father. Although the job is very hard, the market gardener becomes attached to it."[6]

Other books followed, through the First World War and beyond, which allow us to trace the evolution of the vegetable producers' trade in the heart of a great capital, in times that seem very recent but also very distant, given how much lifestyles have changed.[7]

Parisian market gardeners are called *jardiniers-maraîchers* in French, which contains the word *marais*, meaning "swamp" or "marsh." It is a reminder of the time long ago when vegetable crops were grown in wetlands, in the small spaces left free by urbanization, when regulations were much looser than today. In 1845, food crops within the walls of Paris covered about 1,378 hectares (3,405 acres), divided into eighteen hundred gardens, each about 7,650 square meters (1.9 acres). They employed nine thousand people or five people on average per garden: the master gardener and his wife, the day laborers, and a hired boy or girl, more often than not children.[8] The work of these nine thousand market gardeners was enough to supply the city with vegetables.

Increasing urbanization raised the price of real estate, gradually pushing market gardeners to the periphery, inside and outside of the city limits, the border of which was repeatedly redesigned. The producers plying their trade in the city were penalized by the cost of land and subjected to competition from market gardeners who settled in the country, sometimes very far from Paris, where the cost of land was much lower. To remain economically competitive, Parisian market growers were constantly driven to improve their techniques, which they did by focusing on two things: producing all year and producing more per unit area. They could offer vegetables during the winter and early spring, while their colleagues outside the walls had nothing to sell, and they reaped better returns on every inch of land. Their level of expertise, brought on by constraints, was unmatched.

How did they manage to achieve these results, and how might this interest us as gardeners and market farmers of the third millennium? There is certainly much to learn from reading their textbooks.

Create Soil

Market gardeners took extremely good care of their land. One account calls them the "goldsmiths of the ground." They had to create their soil, since what existed did not fit their criteria. To do this they used the virtually unlimited amounts of manure in Paris, where animal-powered transportation was the standard at the time. By composting large amounts of manure in the heart of the city, they rendered a great service to the metropolis, and illustrated two of the principles of permaculture: Any product of a system that is not reused inside the system becomes a pollutant outside; and waste from one must become the resource of another.

Creating soil was a long process: "When a market gardener established a new field in the marais that had never been cultivated, it took him a few years to make the ground amenable for growing crops. During the first years, it took considerable amounts of fertilizer to make the land productive."[9] It was a great loss for a market gardener to be forced to leave his fields, driven by urbanization, to start over somewhere else.

It should be stressed that this approach was almost forgotten with the advent of mechanization and industrial fertilizers — chemical or organic. As we have mentioned, in practice creating soil is only possible in small areas, and subject to the availability of organic matter resources. When market gardeners and large-scale agricultural producers have benefited from mechanized gear, allowing one person to work large areas, it didn't lead to "progress" on all fronts. From the soil standpoint, mechanization is a regression, thanks to the erosion and disintegration it generates. It has become technically almost impossible to rejuvenate the soil of these surfaces, due to their size and the amount of organic matter needed. Creating topsoil has also been regarded as less of a necessity due to soluble fertilizers. Mechanized farmers have gradually stopped feeding the soil that nourishes plants and started feeding plants directly with soluble fertilizer. The result, over time, is absolutely not comparable because the first approach creates humus, while the second destroys it.

Producing All Year

In their effort to feed the city with vegetables produced locally twelve months of the year, Parisian market gardeners demonstrated admirable inventiveness. They used manure resources, super-abundant, to generate heat in addition to fertility, through the hotbed system. All year, the cart that left seven days a week in the middle of the night to deliver vegetables "à la halle" brought a load of manure back to the growing area. In the fall, the market gardeners began creating their hotbeds, mixing fresh manure with manure that had already heated up from the natural composting process in their well-managed manure heaps. They carefully stacked the layers of manure in trenches or on the ground in raised beds. Glass frames or cloches — each gardener possessed significant numbers — were then placed on top of the heated manure. The seedlings were planted in the nursery and then transplanted usually twice, to optimize space. Heat from the decomposing manure allowed vegetation to grow even in the heart of winter, with incredible effort by the market gardeners who watched, day and night, to protect their crops from frost. If the heat diminished, reheats — piles of fresh manure — were mounted around the cloches and glass frames. These were covered, at night or in cold weather, with mats made of rye straw, sometimes two or three layers of them, made by the market gardeners themselves during their morning or evening candlelight vigils. There was no time for such tasks in the daylight hours.

The constraints of handling thousands of glass cloches and hundreds of frames, alternately letting air in and then putting the covers back to maintain temperatures, seem unworkable in our time: "The market gardener, for seven months of the year, worked eighteen to twenty hours per twenty-four, and during the remaining five months, those of winter, he worked fourteen and sixteen hours a day, and often had to get up at night to check the thermometer and double the straw covering the bells and frame containing his hopes and his future, which one degree of frost would annihilate."[10]

The decomposed manure transformed into fertilizer for the soil. Generous additions of this compost were spread around the bed, as was straw, improving the fertility and supporting the enormous production.

Crop Associations

As is so often the case, the success of the Parisian market growers can be explained as the implementation of a set of complementary strategies converging toward a common goal. Crop associations were part of these strategies; they are particularly interesting for our time. Our colleagues in the nineteenth century, in their effort to pull the most out of every square meter cultivated, elevated vegetable growing to an art form. The descriptions of their crops contain valuable information on plant associations that were proven effective at the time: Lettuce was grown with carrots, radishes, spinach, or parsley, and then replaced with cauliflower. As in our own gardens today, the idea was to interplant the next crop before the previous one had been completely harvested. A definition of this technique, taken from the principal reference book of the time: "We call intercropping the art of planting in an area already occupied by vegetables whose growth is much faster. This practice is common among market-gardeners, who want to take full advantage of their land . . . Thus, in a bed of artichokes planted in the fall, we plant beans or cabbage between the rows. If we plant artichokes in the spring, we put potatoes or romaine lettuce between the rows, and so on. Practiced in a well-cultivated garden, this method doubles the production; but it becomes counter-productive in a garden that is not well maintained."[11]

Applications at La Ferme du Bec Hellouin

The landscape has certainly changed dramatically since the days when Paris was populated with a variety of food gardens. Yet in many ways, we will probably return to a situation that is not without analogies to the market gardening days of yore. The constraints posed by the economic, ecological, and social crises into which our society is sinking will lead us to reconsider in depth the agricultural techniques that prevailed in past decades.

Mechanization, linked to the availability of cheap and abundant fossil fuel, has caused highly efficient manual techniques to completely disappear. It is undeniable that the quantity of food produced, per cultivated meter and per calorie invested, was incomparably higher in the "marais" of 1850 Paris than in market farms in 2014. This efficiency in manual practices was not recognized when the agricultural world was turning toward mechanization, as the machine cut the work of the small farmer tenfold and lightened his burden.

The success of the market gardeners, who assured Paris exceptional food security, can become a source of inspiration for tomorrow. In the coming years, it will become necessary to provide the needs of our communities as locally as possible, because food will no longer travel from one end of the planet to the other. But how do we relocalize food production in the heart of cities? There is no room for tractors there, as agricultural land is becoming far too rare and expensive. Only forms of intensive organic gardening are practical in the city. Thus, the heritage of market gardeners puts us on the path to possible solutions. As I. Ponce claimed in 1869, it is possible to "provide vegetables for 1,000 individuals by cultivating an area that once did not even nourish 50."[12] Feeding a thousand people on a hand-kerchief — is this not the dream of many elected officials? It was possible 150 years ago, and much more so today, because we have notable advantages over our predecessors. At La Ferme du Bec Hellouin, many techniques of the market gardeners of long ago have inspired us.

Taking Care of the Soil

We are confronted with soil quality on a daily basis at La Ferme du Bec Hellouin, because of the poverty of the substrate on which we conduct our beloved market gardening work. Microagriculture gives us the power to build up the soil, to create humus.

And we have faced a key question: In the initial phase of soil creation, how much organic matter should we import? Current legislation limits nitrogen addition to the acceptable thresholds for groundwater.[13] But it is possible to input high levels of organic matter, punctually and on very small surfaces, in compliance with the legislation. We've found that the question should be addressed on a case-by-case basis.

What organic matter is locally available, especially outside of rural areas? The horses' hooves no longer echo in the arteries of our cities! But the high concentration of its population offers plenty of kitchen waste to compost, not to mention the "humanure" so dear to the Anglo-Saxons, which has value even if it is not directly spread in market gardens. Composting part of the waste of the city, in place, would be closing the organic matter loop. With a little imagination, the solution is simple; composting is an easy and practical activity that can even be done at the individual or family level.

Growing in All Seasons

The technical advances for four-season growing have been considerable. Greenhouses and plastic tunnels have advantageously replaced the heavy glass frames and cloches of our ancestors. Lightweight, thermal, and easy-to-handle material used today in gardening can create a microclimate by trapping heat from the ground. Using a thermal cover within a greenhouse gives crops quite effective double protection from frost, even in extreme cold.[14] We also have far more choice in vegetable varieties than our predecessors did, and that larger selection includes, notably, vegetables well suited to winter conditions.

The hotbed technique has almost completely disappeared; you'd have to search long and hard for professional producers who still use it. Creating the layers requires intense work, while it is so simple to use an electric heating mat to achieve late-winter planting, or use gas heating to keep a greenhouse frost free.

Yes, but . . . Electricity and gas have an impact on the climate. Could the hotbed method become an alternative to fossil fuels, and also be economically viable? We have been testing hotbeds in our greenhouses since 2013. The heating manure, which reaches 104 to 158 degrees Fahrenheit (40 to 70 degrees Celsius) in three days, allows us to raise our first seedlings in the nursery. We harvest radishes and young shoots in January; the peas are covered with flowers by mid-February. When the seedlings are transplanted, we replant seedlings of high-value crops (like tomatoes, cucumbers, eggplant, and peppers) on the layers, now cooled. We observe that these plants, which benefit from a very rich substrate, mature earlier and are more productive than those that are planted on "normal" beds. In the fall, when these crops have run their course, we have a good supply of compost in the heart of the greenhouse, thanks to the decomposed manure that will enrich the surrounding beds. We are extremely pleased with the results of the hotbed technique.

Combine the Advantages

Certainly, making the hotbeds represents a considerable amount of labor. But I would strongly recommend the approach, which has a payoff that is at least threefold:

- The first seedlings grow early and without the use of fossil fuels.
- Summer crops mature earlier and are more productive.
- A significant amount of compost is produced in the heart of the greenhouse that will enrich the soil.

These considerations illustrate the issues we face when developing a fully manual system of agriculture. It's about keeping up with economic realities. Labor is expensive. Often we find that our practices are time-intensive. But they prove, equally often, beneficial on many levels, and their effects pay off in the medium and long term. A single effort that performs several functions is a winner on several levels. The hotbed gives us these three payoffs, but also improves fertility, benefiting the productivity of the farm over time. Regardless of economic considerations, the satisfaction of avoiding fossil fuel is enough for us to consider this a valid approach.

Likewise, we lay straw or mulch around our crops, as did market gardeners in the nineteenth century. Cutting nettles, picking ferns or reeds, collecting leaves to arrange them between plants is a very time-consuming task. But it is beneficial on many levels, because mulching serves different functions. It protects the soil from direct sunlight, controls weeds, limits the evaporation of water, and fertilizes the soil as it decomposes. If we were to add up the working time saved, from beginning to end, through mulching, weeding, watering, and compost production (which is no longer needed on these plots), we would likely arrive at a total greater than the time it took to mulch. And we have the satisfaction of seeing that, through mulching, our soil becomes more fertile and alive over the years.

Natural farming dispenses with the brute power from burning oil, which is usually accompanied by an immediate benefit but proves harmful in the long term; a bio-inspired approach is slower and gentler, combining various strategies that together make a real difference. This is a complex system, more difficult but also more interesting to undertake.

High-Density Planting and Mixed Crops

Making the choice to do without mechanization also enables us to mix crops and plant them densely, like the Parisian market growers did. *A machine is unable to simultaneously treat two, three, or four vegetable types grown together. The human hand can.*

I confess to a personal inclination to densely pack crops, perhaps too much. The small size of our gardens makes it tempting to overdo it. When we discovered the work of Eliot Coleman and, through him, Parisian market gardening, we tried to blend the two approaches. Why not? With

Coleman's wonderful six-row planter, we sow for example, in a round trip, twelve rows of carrots in a bed that is 80 centimeters (32 inches) wide — one row every 6.5 centimeters (2.5 inches). Reproducing that proven combination of the nineteenth century, we sow, over the carrots, twelve rows of radishes (less densely than if they were alone). In the middle of the bed, we can transplant a row of young lettuce shoots. Radishes quickly emerge and provide a valuable shade for carrots, which enjoy a cool and humid microclimate, while limiting weeds because all the space is occupied. After five to six weeks, we have a normal radish crop. After about seven weeks, lettuce is ready for harvest. We can then replace it with cauliflower sprouts, for example, or fast-growing cabbage. The carrots are harvested over time because the dense planting allows us to collect young carrots to sell in bunches, which lightens the crop, benefiting the remaining carrots that continue to grow. Thus, where a tractor could not grow more than three rows of carrots, we have grown twenty-four rows of vegetables, plus salads and sprouts. The value created per square meter becomes significant. It seems like this cross between two methods has not been tested before. We venture into the unknown. Mixed croppings are dependent on soil and climate conditions. The results can vary from one locality to another. It is necessary to learn the general rules governing the right associations, and then test those suitable for each garden bed.

The Advantage of Verticality

One of the observations that permaculture has drawn from ecosystems is that, in nature, plant systems generally grow in a stepped manner, thereby getting the most from the luminous energy of the sun. Yet modern agriculture emphasizes "flat" systems, in two dimensions only, which go against natural laws. Marc Dufumier, an agronomist well known for his commitment to advancing agroecology, hammered this point into us at a training session on the farm: "Let not a ray of the sun fall to the ground without being caught by a leaf!" Now strive to create multi-tiered plant structures.

Playing with verticality makes more efficient use of space above and below the dirt. The stems and leaves are layered upward in the air, but it is equally important to consider root layers underground, where all of the root space is occupied. Plants with deep rooting act as mineral pumps for plants with shallow roots. The soil is densely inhabited by root-balls, which we try as much as possible to leave in the ground at harvesttime. They naturally enrich the soil with organic matter. Soil life intensifies, especially mycorrhiza. Mycorrhizal fungi live in symbiosis with the roots; their role in plant nutrition is important and becoming better known. Mixing plant species also allows growing vegetables with peas and other members of the bean family, which have natural nitrogen-fixing properties thanks to their symbiotic bacteria, benefiting their companion crops. Another advantage of polyculture is that the threat of diseases and pests is naturally contained.

For now, with just eight years to reflect upon, we're only at the beginning of this experimental approach to mixed cropping. Sometimes it works, but not always: The results can be disappointing. We have learned that we must avoid associating two long-growing crops like tomatoes and carrots, peas or beans, for example. But each year, the list of successful mixed crops grows. We would not turn back for the world.

Our crop mixings are most numerous, and sometimes complex, during summer in the greenhouse. The economic results show that they are the most "successful" plots of the greenhouse. Six to eight crops grown this way can yield more than 200 euros (about \$216) per square meter per year — even if all the crops aren't successful.[15]

Exploring New Avenues

Is the advice from the market gardeners of so long ago really pertinent? Is it still relevant to choose the smallest plot of land and cultivate it exceptionally well? Can their model provide an alternative to the general impoverishment of agricultural land, to the health declines springing from our modern diet? Caring for crops intensively and manually is a constant refrain that we have found throughout our investigation, in Paris, in California with Jeavons, or in Maine with Coleman. It seems possible to integrate nineteenth-century market gardening methods more fully into an approach suitable for small areas — one that is both more environmentally friendly and cost-efficient. Still, it is not enough for us to make a copy-and-paste model of nineteenth-century market gardeners. We have to adapt it more effectively to our context, in our time.

This point may seem simple, filled with sound reasoning. But it was, for us, one of the most difficult to integrate. It required us to rethink the mental image we had of organic gardening. At our farm, we had taken for granted the official data that advocates a surface of at least 1 hectare per market gardener. Once again, we have seen that the main obstacle to the adoption of new practices is within us, in our interpretation of the world. Innovation requires a break from old assumptions. *Pense hors de la boîte*, say the French: Think outside the box. This requires us to open a space in our imagination, a clearing in the dense forest of our interpretations, so that something new can happen.

A degree of bewilderment and naïveté is also helpful when venturing beyond the well-marked trails.

TWELVE

Exotic Influence

Practicing non-action means working in inaction, tasting that which is tasteless, growing that which is small, building up the little, responding to offenses by virtue, elaborating on the difficult with the easy, doing great things with what is tenuous. In the universe the difficult things must be done by the simple. Great things can be accomplished by the imperceptible.

Lao Tseu (Laozi), in the Tao Te Ching

*F*armers in Asia were, for four thousand years, champions of agriculture. We have much to learn from them. Perrine, having lived and worked in Japan and China for six years, has a deep affinity for the Far East. Her connections have continued over the years.

"Just Enough": The Teachings of Traditional Japan

Japan had a profound ecological crisis in the sixteenth century. Its population increased sharply, jeopardizing the supply of the island environment's natural resources, with little arable land to compensate. Population explosion, deforestation, erosion, degradation of coastal waters, and a growing conflict of interest between cities and the countryside had plunged the nation into a crisis that seemed hopeless, a deep systemic meltdown reminiscent of our current situation. While elsewhere in the world brilliant civilizations have not survived such shocks, a remarkable reaction helped rebuild Japanese civilization to a sustainable level during the late Edo period (1603–1868), until the time of Japan's influence by Western industrialization.

At the beginning of the Edo period, almost all of Japan's cultivatable land was farmed, feeding just twelve million people. These lands were, for the most, depleted. Two hundred years later, after a period of widespread ecological restoration, these same areas largely fed thirty million people. Deforestation had been controlled and trees replanted. The land had regained its fertility. At all levels, the actors of society were cooperating in order to find the right balance between the needs of humans and the resources that the islands had to offer. The standard of living had improved, the Japanese were well fed, decently housed and clothed, their level of health was good. It was a result perhaps unequaled elsewhere, then or now.

These developments were the consequence of good governance and technical progress in agriculture — hydrology in particular. "But more than anything else, this success was due to a pervasive mentality that propelled all other mechanisms of improvement," explains Azby Brown in his book *Just Enough*. "This mentality drew on an understanding of the functioning and inherent limits of natural systems. It encouraged humility, considered waste taboo, suggested cooperative solutions, and found meaning and satisfaction in a beautiful life in which the individual took just enough from the world and not more." The just-enough mind-set, writes Brown, "guided the daily lives of millions of people in all sectors of society."[1]

The Pioneers of Natural Farming in Japan

The sobriety that helped rebuild Japanese society challenges us: Will we show the same wisdom when everything leads us to frenetic consumption? There was no hype in the Japan of the seventeenth century.

The shock was violent when Japanese society moved from the Edo era to radically different Western modes of production and consumption. We understand why some charismatic figures emerged during the twentieth century, rising against the industrialization of agriculture and advocating a return to the values of equilibrium and respect.

Virtually unknown in France, the naturalist and agriculturalist Mokichi Okada (1882–1955) developed a natural agriculture practice based on a deep respect for life, logically banning the use of synthetics. He founded Shumei, an environmental but also a spiritual organization particularly interested in links between farmers and consumers. Okada, from Tokyo, had grown up in extreme poverty and suffered from poor health. His life was a series of challenging tests, hence his desire to build on core values. He saw agriculture as an art and believed that the relationship we build with soil and crops has the power to change our lives. Shumei offers a deep connection, physical and spiritual, between humanity and nature, and invites us to respect its integrity, which requires an understanding of the laws of equilibrium, of harmony, and of the interactions that govern it.

"There is a connection between the simple acts of growing and consuming our food and the larger global problem of cultivating a more peaceful world," Okada wrote.[2] According to Shumei, farmers can be the architects of a fruitful reconnection between their contemporaries and nature, the seasons, and people's bodies and health through the healthy food they produce. In this country that gave birth to the first forms of community-supported agriculture, or CSAs (called *teikei* in Japan), Shumei campaigned for a relationship based on mutual recognition: consumer gratitude for farmers, whose work provides them with healthier products, and appreciation of farmers toward consumers, whose long-term commitment allows them to live in dignity in their profession.[3] His message, however, spread more widely abroad, where it carried fewer religious connotations. The renowned Rodale Institute is testing his techniques in the United States.

A Barefoot Revolutionary

Another Japanese pioneer who acheived worldwide fame is Masanobu Fukuoka. His book *The One-Straw Revolution* shaped a generation of readers.[4] Like Okada, Fukuoka opposed the productivist drift of post–World War II Japan. He integrated his ideas on natural farming with a broader perspective, a reflection about the future of our world and spiritual considerations. He is regarded as one of the founders of organic farming, with the Austrian Rudolf Steiner, the Swiss Hans Müller, the English Lady Eve Balfour, and the American Jerome Irving Rodale.

Fukuoka advocates a natural farming method based on nonaction. How to understand this concept? Some of his European followers misinterpreted it; the result was a spectacle of overgrown gardens with very low productivity, which discredited natural farming for many agricultural stakeholders.

However, according to his disciples, Fukuoka worked hard. Reviving and improving traditional methods, he succeeded, after decades of research, in developing sustainable rice fields as productive as the most intensive rice farms in Japan.

Nonaction agriculture refers explicitly to the nonaction underlying Taoist thought. It is a concept that is difficult to grasp for us Westerners, who strongly favor action in our relationship with the world. Failure to act, as I see it, does not mean the absence of action, but rather priority given to the search for a well-adjusted temperament from within. If harmony and serenity reign in the microcosm that we are, then we are in tune with the powerful energies at work in the cosmos because all beings are inhabited by the same energy. Therefore, our mere presence affects these forces, in subtle ways. The order that we have established within spreads to the world around us. We generate, with minimal effort, a positive transformation of our environment. Attaining this righteous state can only be achieved through long-term practice of meditation, listening, and a profound acceptance of what is. With this, we embody the paradox: working through inaction, accomplishing great things through the imperceptible.

Applied to agriculture, this concept is an invitation to:

- Enter into a profound observation of the surrounding nature to intimately understand the forces at work — focusing on slowness, respect, attention.
- Avoid doing harm, weakening the vital potential of the medium. A number of common farming practices need reconsideration, from tilling to the application of toxic substances. This means we need to aim to do less, rather than more.
- Carry out only the indispensable actions that go in the direction of life.

Bokashi (ぼかし) and Other Magic Potions

Was it Taoist spirituality that prompted Japanese farmers to be interested, for thousands of years, in forms of soil fertility that are invisible and yet so terribly powerful? Centuries before the invention of the microscope and the discovery of bacterial life, the peasants realized, at home, the usefulness of microorganisms in soil cultures, true "magic potions." By stimulating the microbial life of their lands, they played, in subtle and elegant ways, with the invisible forces that govern the great alchemy of fertility. When it is known that a soup spoon of healthy soil contains up to one thousand billion microorganisms, we understand the effectiveness of an approach that promotes the multiplication of these invisible actors.

In 2010, Perrine returned to Japan, invited to participate in a symposium on CSA (community-supported agriculture — *teikei* in Japanese. It was an opportunity for her to stay — too briefly — with Japanese peasants. She was fascinated by the quality of the land, the blackness of their small, lovingly cared-for fields, and, found on each farm, the Bokashi — a concoction of various types of domestic plant or animal waste made into highly efficient concentrated fertilizer of liquid bacteria.

Cultivating useful microorganisms in the soil deeply struck Perrine's imagination. The idea made so much sense, and she passionately went down the Bokashi road, forging ahead assiduously ever since. Her approach is essentially intuitive. Even though she reads many scientific publications, we do not have the correct measuring devices to verify the exact mix of her liquid batches. What intrigued her first was the idea that Bokashi might help us solve some problems. Like most farmers growing crops on raised beds, we suffer from an overabundance of slugs. These are scavengers, super-prosperous because beds offer them both shelter (comfortable mulch) and cover (an abundance of organic matter, decomposed a bit). Perrine thought that if we could decompose this organic matter more quickly, we could rid our soil of this slug problem. In addition, soil analysis at the farm has highlighted the fact that nutrients are partly blocked by excess calcium. Increasing the population of microorganisms that make nutrients bioready could be an exciting avenue to explore.

Fermentation or Decomposition?

Back from Japan, Perrine began to make Bokashi at La Ferme du Bec Hellouin, but once again, at the time, literature on this subject was rare. In a fun moment of serendipity — they are frequent on the farm — a prominent Japanese agronomist came to spend a day with us the following week. Mr.

Kawai is a renowned entrepreneur who invented the presown seed mat. At lunch, we asked him if he knew anything about Bokashi. Kawai laughed loudly before answering us (in Japanese, of course): "Dear friends, I created the first Bokashi factory in Japan." There followed a long correspondence in which he advised Perrine on the selection of useful strains.

In Japanese, *Bokashi* means "fermented organic matter." This type of preparation is inherently different from composting: The latter implies aerobic decomposition (in the presence of oxygen) of the organic material, while the latter, Bokashi, implies anaerobic fermentation (devoid of oxygen). The advantages of fermentation are many: speed, the absence of greenhouse gas release, nutrient concentration.[5] Once incorporated into the soil, the Bokashi will develop the good bacteria favorable to rich, well-structured, and healthy soil. It also has a decontaminating effect. Enthusiasts generally consider Bokashi fermentation to be nobler than decomposition, and some even think that massive compost heaps are "pure" rottenness! Perrine, happily, does not go to these extremes, and we are together finding ways to improve our techniques by studying these exotic influences and adapting them for our own use.

Soil microbiology is very complex and, in part, poorly understood. For us, gardeners and farmers, some simple benchmarks are invaluable. It is possible to classify microorganisms in three categories:[6]

- **Dominant-negatives** are decay agents. They promote diseases and represent about 10 percent of all microorganisms.
- **Dominant-positives** have regenerative effect. They feed plants, strengthening their immune systems and improving their resistance to diseases and parasites. They also represent 10 percent of microorganisms.
- **Neutrals** are opportunists: They imitate the positive and negative microorganisms, aligning their action with that of the dominant group. They form the remaining 80 percent.

Obviously we want to reinforce the dominant-positives: When we do, the results throughout the microbial population are powerful. When the soil is inhabited with what's called good bacteria, there is no room for pathogens and parasites.

The anaerobic fermentation process of Bokashi is similar to that used for the lactic fermentation (or lacto-fermentation) used in the production of well-known foods: sauerkraut, sourdough, and vinegar in the West, but also kimchi from Korea, natto and tofu from Japan, Indonesian tempeh, kefir, kombucha, and many others. These fermented foods have a positive effect on intestinal flora and digestion.

Bokashi is interesting on many levels. This efficient preparation is relatively simple to manufacture, and all small farmers, wherever they are, can produce it on their farms by taking advantage of domestic and natural waste, without the need to use any energy source.

Here is a free fertilizer, self-produced, not dependent on fossil fuels, and thus decentralized to the extreme. Bokashi enriches the soil and ensures the independence of the farmer, who is master of his or her own fertilizer production. This is the exact opposite of the industrial approach to fertility. Hidden behind the choice of fertilizer type are, therefore, radically opposing viewpoints.

A Kitchen for the Earth

Perrine, for three years, has launched into multiple Bokashi preparations, and the house is always filled with jars and cans with uncertain contents, sometimes surprisingly fragrant. My favorite farmer is also passionate about culinary preparations using microorganisms and lacto-fermentation, very beneficial to our health. She loves kimchi, the Korean dish that has not yet, admittedly, seduced the rest of the family. Even though we jokingly call her fermentation experiments her witch's brews, it is fascinating to see my wife prepare, with so much care and love, a real kitchen for and from the soil. She nourishes our family with lively dishes, made with the same principle she uses to nourish the soil. We cannot, without a scientific investigation, measure the contents of microorganisms in Perrine's Bokashi. However, we did sow beet seeds in identical trays, and enriched the soil of one tray with biochar charcoal chips soaked in a solution of microorganisms. These chips are created by pyrolysis of wood and enhance soil life by storing water and oxygen and offering shelter to soil microorganisms. After three weeks, the tray enriched with microorganisms was strikingly different: Seedlings were two to three times taller than those in the other tray.

In recent years there are various fertilizers commercially available, liquid or granular, that make use of effective microorganisms (EMs). EM fertilizer is permitted by European regulations for organic agriculture.

Korean Natural Farming

There is a great satisfaction in making one's own Bokashi batches, but in Europe we do not have the knowledge of Japanese farmers or the same raw materials they use (animal blood, fish waste, and so on). To start, it's generally necessary to purchase the efficient strains of microorganisms. Perrine's research revealed that buying these commercial preparations, sometimes rich with sixty different bacterial strains, is not necessarily efficient, because many transplants die, unable to adapt to the new soil condition. The ideal solution, therefore, would be to work with the local bacteria, culturing them to multiply before returning them to the ground. Perrine then sought to master the whole process. She discovered an interesting approach, born in Korea forty years ago: Korean Natural Farming, developed by Cho Han Kyu. This method is used with excellent results in parts of Asia and Hawaii but remains almost unknown elsewhere.

Dr. Cho advocates gathering local microorganisms from various parts of the farm in order to obtain a large number of local strains of indigenous microorganisms (IMO). Effective microorganism cultures are small colonies of bacteria, fungi, and protozoans, very economical to implement since no purchase is necessary, and whose use has stimulated the productivity of crops and livestock. "What you need is already around you," says Cho. Unlike microorganisms purchased on the market, those adapted to the local environment will survive changing climatic conditions. Fermented plant juice is also used, especially that containing the growth hormones concentrated in the young side shoots of plants in spring (in the sucker leaves of tomato plants, for example).

After two years of studying Korean Natural Farming, Perrine concluded that the protocols prescribed to achieve microorganism cultures are complex, perhaps excessively. In addition, they are formulated for tropical climates. She decided it must be possible to draw on these diverse approaches to develop batches with local indigenous strains. Her research is now going in this direction. Cultivating her own strains of bacteria and yeast, she reaches an interesting level of fermentation in just a few days. Moreover, to make bread for the family, she no longer uses yeast. She replaces it with a glass of fermented liquid from her batch: The bread is light, sweet, and has a very pleasant taste.

Now we feed ourselves with the same good bacteria as our soil! The idea pleases us very much.

Terra Preta: The Indian's Black Soil

At La Ferme du Bec Hellouin, we are as passionate about terra preta as we are about Bokashi. In Brazil, the phrase literally means "black earth." In the nineteenth century scientists were astonished by the discovery, in different parts of the Amazonian basin, of very black soil spots with astonishing levels of fertility, in areas ranging from a few square meters to several hectares. This black soil could produce substantial harvests for several decades without weakening. It contradicts our stereotypes of Amazonian land: poor laterite soils, leached by heavy rains, unsuitable for agriculture. A question then arose: Had the terra preta areas previously supported intense agriculture, and therefore large populations? But the Indians remaining in the nineteenth century formed only scattered groups of low density. Had the Spanish conquistadores been right? During their first descent of the Amazon River, they had described densely populated cities, but they were not believed. The presence of terra preta seemed to attest to the existence of dense Amerindian populations, which did not survive the encounter with whites.

Scientific study of this terra preta shows that it is indeed anthroposol — soil created by humans. It contains a high carbon content, gained from wood and vegetation burned during the harvests, but also remnants from the combustion of domestic hearths. This black, ancient soil dates back to pre-Columbus times and forms a layer that measures up to several meters thick![7]

Its extraordinary fertility comes from its charcoal content, which also gives it its color. In itself the charcoal is not a fertilizer, but it contains an extremely large number of microcavities that significantly improve the ability of soil to retain water and guard against leaching because these microcavities are all niches for bacteria and nutrients.

In recent years enriching soil with charcoal has proven successful, particularly in the tropics, to restore degraded land. Charcoal manufactured for this use was baptized into the English lexicon as biochar ("organic charcoal"). Its manufacture is a simple process of burning, at low temperatures, organic materials in the absence of oxygen (pyrolysis). These materials may be wood or usefully recovered waste products such as crushed sugarcane or rice hulls. Biochar is then incorporated into the soil. It is not degraded by the microbial life of the soil and can remain in place for centuries, like the terra preta found in the Amazon.

From the Amazon to Le Bec Hellouin

Manufacturing biochar does not necessarily require sophisticated equipment. On the farm we began with large barrels, before ordering from our friend Vincent Legris, Fabriculture, a biochar boiler.[8] Wood for biochar is enclosed in a metal barrel (in this case, a 200-liter/55-gallon metal drum); combustion is initiated and ignites the wood in a small section surrounding the barrel, itself positioned in the boiler. After a while the wood of the inner container, heated in the absence of oxygen, emits gases that escape from the can through holes (indispensable, these holes; without them everything would explode). These gases are burned in the boiler, once recovered after combustion. It seems important to manufacture the biochar in a double-wall system, as described, allowing the combustion of these gases, as the simplest systems throw the gases into the atmosphere, where they contribute to global warming—which goes against the intent of using biochar.

After a few hours of combustion and overnight cooling, the pyrolysis-charcoal is recovered from the inner shaft and can feed the forge or barbecue, or be pulverized and incorporated into the soil after being soaked in a solution of microorganisms. We also run tests with INRA (France's National Institute for Agricultural Research), using biochar manufactured in Italy.

Sticking to permaculture logic, we do not want to lose all of the heat from the combustion of wood. Remember: Each item must fulfill several functions. So we intend to transform our boiler into a masonry heater and install it in the greenhouse. We will make our biochar during the cold nights of spring, when the summer crops that are susceptible to frost have been planted.

As you can see, we amuse ourselves frequently by testing various ideas originating in the four corners of the world. Exploring new ways to approach agriculture is emotionally and intellectually stimulating. Every evening, when the children have gone to bed, Perrine sits in front of her computer as I immerse myself in books (generational thing!). We each have our pet projects to pursue. *Do you ever take a vacation?* people sometimes ask us. We would love to, but we have no time . . . and, frankly, after a few days, we miss the farm. It's much more enjoyable to do our jobs than to relax.

THIRTEEN

Genesis of a Method

Live simply, so that others may simply live

— *Mahatma Gandhi*

*F*or several years, we have aimed to enrich our practice with various influences discussed in the preceding chapters. We have been thinking about how to organize these influences so that they converge harmoniously toward a common goal: to produce an abundance while regenerating the environment.

The ideal approach does not exist. What matters first and foremost is to devise an agrarian system that is fully adapted to:

- The surrounding environment, taking into account its natural characteristics (soil, climate, water, natural vegetation), but also its societal characteristics (networks, nearby consumers, peculiarities of the local community).
- The creator(s) of the project: What are the aspirations, dreams, competencies, and capacity of the one or ones who create the farm?

What is taking place at La Ferme du Bec Hellouin is nothing more than that: the synthesis that Perrine and I develop, in light of our aspirations, in our own context. The result of our efforts is by no means a model farm, or an example. Our course was chaotic and fraught with countless pitfalls. The list of mistakes is very long.

Reconciling Openness to the Public and Work in the Gardens

La Ferme du Bec Hellouin was quickly identified as one of the places where viable alternative agricultural practices were being developed. While we were still beginners, we began to receive more and more visits and requests. This was an opportunity, as many of these visitors were very skilled, rich from several decades of experience in the agricultural world sometimes, and their perspective on the farm was informative. But it was also a big constraint, because the number of solicitations quickly exceeded what we could reasonably manage. Soon hundreds of visitors, and thousands of telephone calls and emails, poured in, to which our benevolent replies nibbled at our farmwork and family life. We hired a secretary, largely dedicated to meeting these demands, which lightened the load but contributed to our economic deficit. The media were equally interested in the farm.

During those dark years, we were constantly disturbed and could not count on the occasional day of rest. The phone rang at all hours of the day and sometimes at night, even on weekends. People would knock on our door at night: "Good evening, just visiting your farm, where can we camp?" Probably, these visitors did not realize that behind the farm there was not a research office or other solid enterprise, but just a family of farmers, with its limitations and frailties. There were also increasing financial difficulties. We were suffocating on the edge of implosion. We had to mourn the initial dream and accept what life offered us. This did not happen smoothly or without some resistance. But encouragement from many confirmed the need to continue on this path, despite its difficulties.

This encouragement came from wonderful people who supported us and often became good friends. First there were some great pioneers of organic agriculture and agroecology, whose books had put us on the path, who agreed to come voluntarily to teach at the farm: Pierre Rabhi, philosopher and poet farmer; Claude Aubert, author of the first book on organic farming in France, and founder of Living Earth; Marc Dufumier, scientist and brilliant speaker, whom we listened to for days on end. And there was Philippe Desbrosses, one of the founders of organic agriculture, who claimed a special place in our hearts and became a close friend. His book *Nous redeviendrons paysans* (*We Re-Become Farmers*) was particularly useful.

Philippe is a passionate man, deeply spiritual, entirely devoted to the pursuit and dissemination of solutions to build a better world. He founded the Ferme de Sainte Marthe forty years ago, and has amassed an outstanding collection of organic seeds. He launched an organic farming school in Sainte-Marthe, France, that teaches dozens of people every year.[1] Full of humanity and kindness, Philippe took us under his wing and comes regularly to the farm. He is honorary president of the Sylva Institute, which manages our research programs.

The trust and friendship of these great pioneers for the little beginners that we were made me think they were passing, in a way, the baton: It is up to our generation to continue the work they started, supporting it with their advice and their watchful eye. Left alone, Perrine and I are sometimes inclined to doubt what we do. Their positive attitude, and that of so many other people, confirms that there are important reasons to explore permaculture practices.

How Many Salads per Square Meter?

If the farm has survived it is also thanks to the friendship and guidance of François Lemarchand. Creator of Nature et Découvertes (a retailer with stores throughout France, Belguium, Germany, and Switzerland), François is a brilliant entrepreneur and devoted ecologist. He shares his passion for nature and open spaces with his wife, Françoise, creator of the French magazine *Canopée*. Together they are committed to fostering new practices, attitudes, and cultural traits that will shape the emerging world of tomorrow, and helping to realize this with the two foundations they created. Since François does nothing halfway, at sixty-two years old he decided to get agricultural training and did an internship with us. We were in low spirits at the time, and welcoming François every week for several months was a godsend. We needed a co-conspirator to continue to believe in our adventure. His advice was always pertinent. I remember one day in late winter, kneeling side by side, transplanting lettuce in the mandala garden. François yelled at me affectionately: "Your business is managed any which way, you're going straight into a brick wall. You cannot continue to ignore the economic dimension of the farm. In our stores, we absolutely must make a minimum profit per square meter; without that we would put the key under the door." And I saw my dear François transplanting the shoots increasingly closer, the quantities we were reaching were approaching forty per square meter! The lesson was well received. Eliot Coleman, in different terms, said the same thing. François makes it clear that to be copied and adopted by others, an environmentally friendly practice needs to demonstrate its economic viability. It seems he was prepping us for the study that we would launch a few months later.

The farm has had many guardian angels, which has allowed it to overcome more than a few trials. But there was also jealousy and backstabbing, betrayals. The moment we chose to move forward with our plans, we were exposed. An African proverb states: "The more one climbs up the coconut tree, the more it shows his ass!" I understand the irritation that media coverage of a farm created by two new urbanites can arouse in some who have been involved in farm life for a very long time. By accepting media requests (we never invite them ourselves), we get visibility that does not always make new friends for us. So why expose ourselves? Because we think it would be cowardly to live in our cushy corner of paradise, indifferent to the problems of the world, without trying to make a contribution.

First Classes

We had almost no premeditated course in those early years. Things came along, and we had to consider them. We were asked to organize training sessions. There we saw the opportunity to direct this river of solicitations that threatened to shipwreck us. Making ourselves completely available during the defined time ranges of the sessions allowed us to share more. So we have offered training in permaculture gardening and farming since 2009. The sessions took off, as if by storm, filling up sometimes nearly a year in advance. Some trainees participate in four to eight sessions each year, constituting a friendly and motivated team. With nearly four hundred trainees welcomed since the second year, these courses are very quickly becoming a wholly owned business, to learn and manage. We achieved this by relying on the expertise of external trainers who are wonderfully competent in areas where we were weak.

The training sessions alternate theory and practice, and the fact that many concepts of permaculture are illustrated hands-on at the farm attracts many trainees. They can apply the theory in the real world, as they learn it. We believe that teaching gardening and vegetable growing on an organic farm is an asset, because the requirements of production, twelve months a year, put all the teachings to the test of practice. The classes have three different and complementary aspects – production, research, and education. Having all three aspects available at one site is relatively unique in Europe.

The Farm Crew

This fact should be underlined in bold: Well supported from the outside by a developing ecosystem, our farm is built from the inside thanks to a great full-time staff. Over the years our crew has made us proud, approaching their work with investment and seriousness, but also with humor, mutual respect, and kindness, which has allowed us to gradually delegate a number of responsibilities and get our heads above water. We feel very fortunate to be working in an environment of such professionalism and confidence.

It is necessary to be surrounded by quality people to overcome the difficulties of a high-risk business. In 2009 we started, with the usual naïveté, the construction for our visitor center. This project lasted three years and cost double the initial budget. We now have a superb bioclimatic building, entirely made of natural materials — a 400-square-meter structure built in the local tradition of wood frames, filled with stones and mud. The wood was collected from old dilapidated structures, including that of a cider press dating back to 1668. The energy supplied by Enercoop, a co-op distributing electricity from renewable sources, is entirely green.

The cost of this project could have sunk us ten times over. But logic sometimes knows when to yield before the power of a vision. The local and regional authorities have invested heavily in the project. Our region has lagged behind national progress in organic farming, so this training center was well received by Haute-Normandie, the general council of the Eure region, and the agricultural chamber of the Eure, who mobilized support for this investment.[2] And a number of miracles occurred, large and small, to help us make it to the finish line. When we were in an almost hopeless situation, a bank called Credit Cooperatif offered us a large loan. By meeting Jerome Henry, one of its leaders, we (finally!) found a banker — the real deal, passionate about the "human economy," a friend of ecology — and a bank that knows what engagement means. The ecocenter was barely finished and already filled with training sessions, allowing us to honor our commitments and expand the team to face an increasing workload.

It seemed that, like Pupoli, we had positioned our canoe in a favorable vein of current, found our rightful place. During those years we were only trying to discover and apply the concepts of natural farming, yet we were visited by a number of organic agriculture officials from France, agronomists from the United States, including the director of the Rodale Institute, and many others from elsewhere — a Brazilian minister, NGOs, African leaders, permaculturists from Cuba, and more. We were nourished by these warm and open exchanges.

A Farmer Pulled from the Creek

As if the above were not enough, Perrine was then launched into a new adventure. One evening as she waded in the stream, trying to catch wandering ducklings, an elected official from the region, Claude Taleb, called out to her from the bank. He asked her bluntly to shore up the list of environmentalists for the next regional elections. Dripping and surprised, for she had never even contemplated a political post, Perrine agreed in principle. Soon she was elected regional councilor and mandated to develop organic farming in Upper Normandy. Perrine was thrilled about this mission in the field, which put her in contact with the various agencies in charge of agriculture, Safer, and the agricultural schools.[3] Her past experience as a lawyer and organic market farmer earned her some credibility with a diverse set of actors. This commitment and the many connections it has produced broadened her perception of agriculture.

Genesis of a Method

All these expectations were so many invitations to describe what characterizes our approach. We then reflected deeply on how to synthesize the many "good practices" gleaned along the way in a permacultural framework. This led to the Bec Hellouin farm method, which highlights the elements we see as essential to building an economically viable agriculture and permaculture likely to contribute to the regeneration of the environment.[4] This twenty-point method continues to evolve. We personally have not invented anything; our only merit is to have foraged from various sources, then tested and consistently organized multiple approaches. Permaculture is a formidable tool for this.

I was embarrassed and reluctant to baptize this method with the name of our farm, and I still am. But there is merit in linking it to our farm, as it clearly states: "This is what we have achieved in the context of our farm, at the stage where we are in our evolution, with our unique circumstances. This is by no means valid as a set of rules for all places, much less a dogma! But perhaps you can get inspired to build your own method that will allow you to thrive in your context. Our approach can help you avoid mistakes and fatigue, as well as unnecessary costs, and help you get closer to your goal faster!"

It became clear to us, as was mentioned, that the determining factor of success or failure in natural farming is not so much the techniques used (or not used) but our positioning in regard to the biosphere, the mind-set that we have toward market farming. A change of techniques and tools is relatively easy; there are fairs and trade magazines that offer a wide range of choices. But the farmer's job is much more than a set of technologies. The farmer's transformation of the biosphere impacts the landscape and multiple forms of life, for better or for worse. Therefore, it is essential to question ourselves deeply about our positioning toward nature. Are we outside, above the community of the living, or immersed in it, united, guardians of all life-forms that share the same land?

How we shape the landscape we inhabit is a reflection of our inner wisdom. Ecology begins within us; it is spiritual.

It is therefore necessary to make the link between the way we aspire to inhabit the earth and the farm practices that allow us to carry out this vision. One eye on the star that guides us, another on the stones in the road, so we don't fall. And of course we must find the strength to get up after every failure.

FOURTEEN

Launch of a
Research Program

*The human race has reached a crossroads. Our future depends on
the speed with which we understand the extreme seriousness of the
situation.*

— *René Dumont*[1]

A team of agronomists came to spend the weekend of May 1,
2010, with us. There were, among others, François Léger,
director of the research unit SAD-APT (AgroParisTech/
INRA), Stéphane Bellon, head of organic farming for INRA, and Cyril
Girardin, a soil scientist at the Bioemco laboratory. Not knowing any of
these scientists, still being complete beginners in our permaculture
approach, we were in our own small shoes, so to speak, upon their arrival.

After greeting one another, we started to explore the farm. During the
first morning, we just crossed through the island garden. François,
Stéphane, and Cyril were quite obviously interested in the empirical system
we had created by following our intuition. The rest of the weekend was
spent in exciting discussions. Upon leaving us, Stéphane humorously sum-
marized their viewpoint: "We aren't crazy enough to create a system like
this at our research stations; that's why it is so interesting to us!" In response,
we told them that we were very moved, for our part, by their scientific perspective. Later, at a meeting bringing together various practitioners to
create a network of farms experimenting in agroecology, Stéphane said:
"What caught my attention at the Bec Hellouin farm is that Charles and
Perrine produce abundant food on a space that couldn't feed half a cow." We

could have left it at that, stayed in touch with them, especially with François Léger. François has a very sharp mind, constantly navigating the borders of agronomy, sociology, and philosophy. Busy with the responsibility of directing a team of about sixty researchers and teaching the masters class "Environment, Development, Territories, Civilizations" at AgroParisTech, he still finds time to periodically visit the farm. These are always hours of intense exchanges, during which we bombard him with the many questions that arise as we move forward. Many of the ideas and concepts presented in this book have been enlightened by his perspective.

Soon enough we arrived at the idea of a study that would provide a scientific basis for the approach developed at the farm. We wrote the outline defining the project, and François defined the scope of the study:

> The question we would like to address is: "In a system of permaculture gardening without mechanization, what is the size of the growing area required to generate a decent income for a worker who wants to establish and have acceptable working conditions?" This question immediately raises problems related to subjective interpretations of "decent income" or "acceptable working conditions." So we have to rephrase the question with these two issues in mind:
>
> - What is the economic performance that can be obtained in a limited area, arbitrarily set at 1,000 square meters?
> - What is the workload required to achieve this performance on 1,000 square meters and how can it be distributed?
>
> With which we associate a third question, allowing us to address more fully the essential point of sustainability of a market-garden permaculture system:
>
> - What is the environmental performance of this system?

Our agronomist friends thought that a market gardening approach that appeared to be both environmentally beneficial and productive deserved to be studied. By itself, La Ferme du Bec Hellouin, defined by our more or less chaotic trajectory and its particular soil and climate, was not more interesting than any other farm. The goal was to model market gardening permaculture practices that could be reproduced, to help disseminate the methods.

The project proposal was drafted in June 2011 and entrusted to three or four people likely to be supportive.[2] We wanted to work with discretion, in order to stay focused and calm. The dossier was therefore not disseminated online, yet it quickly passed from hand to hand and made the rounds of the small world of agroecology in France, prompting contradictory reactions. Those who knew the farm were generally enthusiastic about the study. But some historical pillars of agroecology jumped to conclusions and called us a host of names because it seemed impossible to create a viable business on such a small surface. It was generally accepted that a 1-hectare (2.5-acre) market garden was the minimum for a full-time salary. They couldn't yet imagine the potential productivity that could be realized with a permaculture approach.

Bertrand Hervieu, a former president of INRA, then president of the environment division of the Fondation de France, came to visit the farm on two occasions and thought that the study deserved to be conducted. The foundation committed, as did the Lemarchand Fondation — focusing on the balance between people and the earth — and The Lea Nature Foundation.[3] A scientific research committee of prominent scholars was formed.[4] In late 2011, the three-year study began.

Modeling a Living System

A cultivated area of 1,000 square meters (0.25 acre) was designated in the heart of our gardens (this means square meters that are *cultivated*, so it doesn't include pathways, composting areas, the buildings), deliberately excluding atypical spaces such as island gardens. This "virtual farm" includes seventy plots. We then proceeded to record everything that goes into these plots (work time by task type, inputs) and all that comes out, down to the last bunch of radishes. Similarly, all the tools and equipment (greenhouses, irrigation systems, and more) necessary to farm were listed and the cost recorded.

We soon faced the difficulty of modeling a living system that is, by definition, complex and evolving — one of François Léger's motivations to start this study. In a conventional agroeconomic study, a crop planted at time A is harvested at time B; the next planting starts at time C; and so on. At La Ferme du Bec Hellouin, because of widespread crop mixing, multiple cropping sequences overlap. Several hundred crops had to be monitored annually as precisely as possible. Many meetings were necessary to define a protocol for collecting and processing data that is sufficiently simple to be usable on a farm, and comprehensive enough to collect all the important parameters.

It is rare that such a complex study takes place within a single farm's production. "The assumption that drives the partnership between the scientists and the operators of La Ferme du Bec Hellouin," François wrote in the project proposal, "is that the results can actually be used to support the production of references. The proposed study aims to produce a knowledge base covering all the dimensions of the production system and its different measures of performance . . . It will serve as a more general reflection on the very nature of references to be produced for 'agro-ecological gardening.' It will help expand the current scientific thinking on agroecology and its principles." Market gardening is a highly constraining activity. The study added another layer of complexity — but also a wonderful opportunity to progress — with the constant vigilance it imposed. The whole team had to be trained to collect data and get used to living with a notebook in the pocket and an eye on the clock. The team expanded as students joined it. The study supported papers and doctoral theses, and now extends to other sites, including La Bourdaisière, in Touraine, largely inspired by our own experience.[5]

Thousands of data points from the first year, compiled daily by the farm team, were fully analyzed and verified three times, under the watchful eye of François.

The Results of the First Year

On July 5, 2013, nineteen months after the start of the study, sixty people gathered in the ecocenter for the presentation of the results from the first year. Surrounding François Léger, Stéphane Bellon, and Philippe Desbrosses was a strong contingent of Belgian researchers, officials from the Eure's Chamber of Agriculture, market gardeners, and elected officials. Sacha Guégan, the engineer in charge of the study on the farm, presented the data and their analysis. They validated our original hypothesis:

- Over a year, the total revenue generated was 32,000 euros (about $36,000), and the associated workload in the gardens was fourteen hundred hours.
- It is therefore possible to support a full-time farmer on 1,000 square meters cultivated using the method of La Ferme du Bec Hellouin.

Some additional findings are worth noting:

- Revenue was determined by assigning an economic value to market garden production. This value was determined by price lists from different sources, primarily one provided by the regional association of organic farmers of Haute-Normandie, giving average prices for the direct sale of organic vegetables in the region.
- The working time, fourteen hundred hours in the gardens, comes out to twenty-nine hours per week. The time spent on site maintenance, sales, and administrative tasks was estimated to be 50 percent of the hours worked in the gardens. So the annual workload added up to twenty-one hundred hours, an average of forty-four hours per week, with four weeks of vacation. This is very reasonable for a farmer, who moreover does not waste time on transportation to get to work.

The team judged these initial results to be very encouraging. The workload was reasonable and probably less than that generated by a large-surface market garden. The profit, even in a mediocre first year, was similar to what is generally accepted as the average results of a hectare of mechanized organic market gardening.

For the year 2013, earnings increased significantly, exceeding 39,000 euros ($44,000) for 1,490 hours in the gardens.

In the beginning of 2014, we were confident in the relevance of our approach and dared to go farther with our densification and crop mixings. By late April that year, some beds in the greenhouse had already generated four plantings and three harvests! Taking a permaculture approach to microagriculture was displaying a productive potential far superior to what we could have imagined at the start of the study, and it has not finished surprising us. This is probably the most productive form of agriculture that exists.

Beyond the numbers, François also asks: What is the quality of life offered by a farm that is, as he puts it, *the extension of a large garden and not the reduction of a mechanized farm*? It is of course impossible to answer a question as subjective as this, but we must recognize that this new form of organic market farming offers a plethora of tasks (we rarely perform the same job for more than an hour at a time) and immerses us in nature, which makes practicing the profession fulfilling and enjoyable.

Room for Considerable Improvement

In the unanimous opinion of the team, these intermediate results (which will need to be verified for precision) are widely improvable. The number of crops grown on each patch ranged from one to eight. The detailed analysis of each of the seventy plots revealed a highly variable margin. Some plots were virtually abandoned. The good economic performance of the best plots was due to the crop mixings and high-density planting only possible with manual labor.

It seemed to us that the adventure was only beginning. On our bio-intensive-agriculture learning curve, we're only in kindergarten! The method used at La Ferme du Bec Hellouin to achieve these results was the fruit of several years of experience. But mechanized agriculture has enjoyed 150 years of research in many countries, with the financial backing of large firms. In terms of natural farming, we still have much to discover — and to rediscover from successful examples of the past. Almost all is likely to be improved, sometimes considerably. In particular, the development of tools adapted for manual microagriculture would lead to gains in efficiency.

A dynamic is afoot. Several study projects have sprung up on different sites. Research conducted at the farm will also be part of a CASDAR (Trust Account for Agricultural and Rural Development) study on agroforestry farming systems, which started in early 2014. François noted, in the second interim report:

> The study launched at La Ferme du Bec Hellouin, and this is undoubtedly one of its main successes, has been the starting

point for a broader set of work that aims to define all the elements needed for a farming method that combines market-gardening and fruit production on very small surface areas. In these systems, the attention to crop layout maximizes biological interactions and allows a significant level of income, thanks to high productivity and very low consumption of inputs and fossil energy. These models, because they require more thought and work than capital, are candidates of the first order to revitalize agriculture in areas which have been excluded (cities and suburbs especially), contributing to the recomposition of food systems by bringing together producers and consumers, and the creation of jobs.

The first results obtained at La Ferme du Bec Hellouin are ultimately very convincing despite a difficult year. We invite everyone to follow and discuss this ongoing work in an equal exchange between researchers and producers.

The results of this first year, once again, should be taken for what they are: a sketch that needs to be clarified. Agronomic research is not built in just one year! We supply these intermediate results only because they indicate that the road is worth exploring.

Permaculture farms have benefits well beyond the food they produce, but only the value of our vegetable production has been analyzed in this study. One can imagine that the other services rendered to society will one day be considered as well. Perhaps then local authorities will support all or part of the costs of a farm's installation, given the positive impact that a market garden farm in town may bring to the community.

The Impact on Biodiversity

The numerical data are only one aspect of the study; they should not overshadow the qualitative dimension of this type of agriculture. Along the way, each year we produce more ecological benefits with this bio-inspired approach. Our meeting with Gauthier Chapelle, from Belgium, has greatly contributed to this awareness. Gauthier has a doctorate in biology and is an agronomist, naturalist, engineer, and consultant, but above all one of the pioneers of biomimicry in Europe.[6] During two visits to the farm, he drew our attention to the biodiversity that had been established in our gardens. Here's what he wrote in the second interim report of the study:

> I was immediately struck by the significant presence and diversity of wildlife on the farm (not to mention the diversity of the domesticated animals). If this biodiversity is in part due to the biotope of this area (running waters, stagnant waters, meadow, brush, wood, etc.), I would like to share two observations that will shed light on my first instinctive impression.
>
> The first observation, of a naturalist, regards birds, and more particularly in this case, a subgroup of passerines, granivores par excellence, that of *Fringilla* (one of the best known is the chaffinch). To my surprise, two days in mid-June was enough time for me to observe seven different types or, theoretically, all possible birds that nest in this part of Normandy.[7] All on a sustainable farm! Wow! And this includes once-common species that are becoming increasingly rare (such as the melodious linnet and bullfinch).[8]
>
> The second observation concerns insects — noting already the presence of transients and several species of dragonflies, including the virgin calopteryx, a testament to the ecological quality of the aquatic areas of the farm. But I was especially taken aback by the abundance of a very special red and black beetle, called the bee-hive beetle; adults were present on many different flowers found in the vegetable gardens and forest-garden, busy feeding and thus pollinating.[9] More significant above all: their larvae are parasites of several different species of solitary bees, which obviously signals a massive presence of the latter on the whole farm. When one knows their role in terms of pollination, complementary to that of domestic honeybees, we can only rejoice.

Note that the presence of all this wildlife on the farm is not a concern: Contrary to popular belief, almost all wild animals are valuable helpers.

The Climate Impact

That same year, 2013, we received a visit from Jean-Marc Jancovici, an eminent specialist in carbon sequestration, along with members of his wonderful Shift Project and Carbone 4 team. We have also had exchanges with our friends from the Pur Project, created by Tristan Lecomte, as well as Jean-Philippe Beau-Douézy, responsible for the reforestation programs of the Yves Rocher Foundation.[10] Discussions with these specialists confirmed our belief that the kind of farming we are doing stores carbon, although this is difficult to quantify, whereas industrial agriculture contributes massively to global warming. Carbon is sequestered in our soil, as it grows richer in organic matter, and in the trees that are ubiquitous in the agroecosystem.

Since global warming is probably the main challenge of the twenty-first century, the fact that agriculture can become one of the levers for containing greenhouse gas emissions within acceptable limits provides some hope.

We were in our seventh season as market farmers in 2013, and we had reached a tipping point. Up to that point, we had tried to do the least possible harm to the planet while practicing our profession. Now, we realized, we could actually regenerate the surrounding environment while keeping our production high. If bio-inspired agriculture became widespread, it would become possible to feed all of humanity while restoring the land on which it operates. This simple realization has given even more meaning to this research, and tremendous energy to move forward.

A Well-Attended Study

In its first year, this study began to be followed by a large number of institutions in the agricultural world and beyond. Perrine was invited to speak at two conferences organized in the European Parliament in Brussels, including Feeding Europe in Times of Crisis, which was also attended by Olivier De Schutter, UN rapporteur on the right to food. I presented the study with François at the first conference on bio-inspired agricultural research, organized by the Ministry of the Environment. Perrine also participated in conferences organized by the Soil Association in England and EcoFarm in California, as well as in exchanges with various people that she met along the way.

We were awarded the Sustainable Agriculture prize in Haute-Normandie, and the Minstry of Agriculture said this when issuing its Hope for Sustainable Agriculture award: "To Perrine and Charles Hervé-Gruyer, of La Ferme du Bec Hellouin in Normandy, for their exceptional effort and innovative experimentation, for their inventiveness and desire to disseminate their findings while continuing to seek out original and promising avenues" — a sign that these institutions are, in fact, open to new paradigms.

Officials from local authorities, major cities, and even European capitals came to the farm to consider the the possibility of starting up microfarms in their region.

There are no words to express our gratitude to François Léger. It takes a heavy dose of audacity to engage a research unit to study an alternative model. Thanks to his skills, and the thoughtful manner that the participating agronomists and their institutions brought to the project, a science-based platform is beginning to be built around a farming method that is at complete odds with dominant agricultural practices.

We hope that an increasing number of farmers will be tempted to explore the potential of bio-inspired agriculture. As more do, and as we all increasingly share our information and experiences, it will not only leverage the impact of this research but will allow us to collectively invent a blueprint for post-oil agriculture.

FIFTEEN

The Forest Garden

*A town is saved, not more by the righteous men in it than by the
woods and swamps that surround it.*

— *Henry David Thoreau, 1851*[1]

*P*errine and I are convinced that the most innovative system we
put in place in our valley is the forest garden.[2] Permacultural
market gardening is the evolution of a practice conducted since
ancient times in our country, while the forest garden is really a new production system in our latitudes — what the English call an edible forest. A forest
that is eaten, can you imagine?

The forest garden concept seems to present a set of exceptional environmental and societal benefits. Once established, the forest garden is a
sustainable system, autonomous, resilient, productive without the use of
fossil fuels, and without water or fertilizer needs. It stores carbon, restores
landscapes, and is a cultivated and wild refuge for biodiversity. In social
terms, the forest garden allows local production of fruits, nuts, berries,
vegetables, herbs, plants for medicines and dyes, mushrooms, wood, and
biomass, all while creating jobs. It requires only small land areas and can
easily be incorporated into a garden.

The forest garden suggests a new form of agriculture that gives trees a
central role.

What Is a Forest Garden?

The forest garden is a form of "wild" agroforestry born in the tropical
regions of Africa and Asia, where some indigenous people live among the

plants that are useful to them, especially fruit trees and berry bushes. These plants form a forest garden providing — in addition to fruit, berries, nuts, and edible leaves — lumber and heating, vegetables, aromatic and medicinal plants, and materials for crafts for use in the village or for sale in local markets. These gardens can also accommodate livestock shelters and beehives.

Here is a definition of this system adapted to our latitudes, from Patrick Whitefield: "A forest garden is a garden modeled on natural woodland. Like a natural woodland, it has three layers of vegetation: trees, shrubs and herbaceous plants. In an edible forest garden the tree layer contains fruit and nut trees, the shrub layer soft fruit and nut bushes, and the ground layer perennial vegetables and herbs. The soil is not dug, and annual vegetables are not normally included unless they can reproduce by self-seeding. It is usually a very diverse garden, containing a wide variety of edible plants."[3]

The Forest Garden, a New Concept in Europe

Forest gardens, popularized in the wake of permaculture, were adapted to latitudes in England by pioneers like Robert Hart and Martin Crawford. It seems that this form of agroforestry with three or more levels of vegetation (Hart describes seven levels) had never before been practiced in Europe, where agroforestry systems were established on only two levels (as with olive groves, meadow orchards, or chestnut trees). They debuted in France only a few years ago and there are still only a few prototypes here, which are poorly understood by both specialists and the general public. They have for the most part, if not entirely, been created by amateurs, and the scientific community seems to have had little or no involvement with the conception or implementation of these prototypes.

It is interesting to note just how profound an innovation the forest garden is, a complete departure from the usual agricultural practices in our Western culture. Our agriculture originated in the Neolithic period in the Fertile Crescent. The plants and animals on which it is based (grain, sheep, goats, and cattle) are from the steppes of Asia Minor. This type of agriculture generally calls for cutting the trees and opening up spaces to create a steppe-like landscape. Since prehistoric times, this deforestation has caused environmental disasters, even contributing to the desertification of North

Africa. But the damage this mode of agriculture causes has amplified since 1950 with the advent of synthetic fertilizers, the development of mechanization, the consolidation of land, and the expansion of the model in many parts of the world. Indeed, the next step after steppe agriculture is desertification, and it is clear that our modern agriculture has a negative impact on soils, water resources, landscapes, and the climate.[4]

But the vegetation that once prevailed in temperate Europe was the forest. If we abandon an agricultural area, it is the forest that will return naturally. Maintaining open space requires constant effort. The central role of the tree is becoming better appreciated. It fulfills a plethora of ecological functions, creates soil, promotes microclimates conducive to life, and stores carbon. It also beautifies the landscape and provides countless human services.

The forest garden is a "concept" from other parts of the world, other cultures, where agricultural traditions have given the tree a central role — as with the first peoples living in subtropical forests. Its development in the West could be a paradigm shift that contributes to remediating the damage caused by the dominant agricultural model, giving back to the tree its essential role in the balance of our agroecosystem.

ECOLOGICAL BENEFITS OF THE FOREST GARDEN

The potential environmental benefits of forest gardens are particularly interesting, and worth summarizing here.

- Judicious use of verticality allows a significant amount of production in a small space. As in tropical forests, a vertical stratum can even be created in the forest garden, with vines climbing up tree trunks.
- The forest garden, once established, is an autonomous system. It requires no mechanical tillage, and hence no fossil fuels or watering.
- The forest garden generates its own fertility. Trees draw up minerals from the soil and return those minerals to the earth in the leaf litter, to the benefit of plants with shallower roots. Some trees are nitrogen fixers. The forest garden does not require fertilizer or compost inputs.

- The forest garden is an oasis of biodiversity, for flora and fauna, wild or cultivated.
- It forms a resilient agroecosystem, since the ability of an ecosystem to withstand the vagaries of the weather is directly related to its biodiversity.
- It is a long-lasting system, whose longevity is at least that of the tall trees that make up the canopy, perhaps one hundred years. The lower levels are replaced when necessary. Surely if we replace the fruit trees when appropriate, a forest garden can have a life as long as a traditional forest.
- The forest garden acts as a windbreak for the benefit of a vegetable garden, for example, and promotes the creation of climates and microclimates favorable to plants, animals, and human beings.
- The forest garden stores carbon and contributes to stabilizing the climate.
- It provides crops without negatively impacting the environment. On the contrary, it contributes to the improvement of soil, water, air, and biodiversity.

THE SOCIAL BENEFITS OF THE FOREST GARDEN

The forest garden also seems to have remarkable potential benefits for humans. Among them, the nutritional potential deserves special attention.

- The forest garden produces food of a very high quality: fruits and berries rich in vitamins and antioxidants, herbs and medicinal plants, and also perennial vegetables and mushrooms. These are the best types of food for humans, those for which we are designed, as they represent the diet that our species lived on for millions of years during its evolution.
- It seems desirable, both for the planet and for our health, to gradually move our diet toward a nutrition based largely on fruits, berries, and vegetables and to rediscover the potential of nuts.
- The proliferation of forest gardens could help promote local access to quality food, creating perennial resources that are

independent of fossil fuels, and thus strengthening the food security of communities.

Like community vegetable gardens, forest gardens can create social ties. Trees and forests have always played an important role for humans, offering fulfillment symbolizing stability.

Toward the Emergence of a New Agricultural Trade — Sylvanier?

Interestingly, the forest garden could prompt the emergence of a new profession in organic farming, one that we propose to name sylvanier (forest gardener). This new occupation has several major advantages.

ECONOMIC ADVANTAGES OF BEING A SYLVANIER

The forest garden mainly produces organic fruits and berries, high-demand food with a high economic value at the market. A small agricultural area could therefore attain a potentially large profit margin. In addition, forest garden products are diversified; most withstand storage (drying, freezing) and processing (syrups, jams, teas), making market production possible throughout the year.

Investment costs (purchasing and planting the land) are relatively low compared with those of a conventional agricultural installation, and amortized over a long time frame. Another argument will resonate with those sensitive to building a legacy: Unlike most production tools, the forest garden gains in value and productivity over the years; sylvaniers who plan their forest garden at thirty may benefit until their death, or decide to sell it to retire for a much higher amount than the initial investment. Trees are a natural form of retirement savings!

The operating costs are low. The tools required for maintaining a forest garden are minimal. Only small equipment, such as a chipper, is needed. Maintenance and harvesting are done by hand. Unlike agricultural methods dependent on annual crops, a forest garden becomes productive gradually. It takes five to seven years for fruit to establish in the upper-canopy trees. This disadvantage, offset by the long life of the forest garden, can be circumvented by cultivating the lower levels more intensively in the early years, when there is still plenty of light shining through. The middle tier starts producing from the second year.

It seems like the net margin and thus the compensation for a sylvanier would be acceptable. The proliferation of forest gardens could promote local economies, creating true value based on the earth and nature, stable and sustainable, unlike the financial and globalized economy that currently predominates.

SOCIAL DIMENSIONS OF BECOMING A SYLVANIER

Inexpensive to install, and requiring only a few thousand square meters of agricultural land, forest gardens can provide a new gateway to the farming profession, which is currently often difficult to break into.

It is particularly interesting to note that the sylvanier profession can be practiced, because of its ease, by a person whose age, situation, or state of health does not allow working the land conventionally. This includes:

- The elderly or those facing early retirement. Fifty-year-olds who become unemployed might find it difficult to return to the workforce. They may, however, be sylvaniers for many years.
- Those who lack the necessary time for market gardening, such as parents staying home to raise families.
- People who lack the physical strength necessary for market gardening, including those with physical disabilities, chronically sore backs, or other ailments, but who are capable of picking the harvest and light maintenance.

By practicing as a sylvanier, many types of people, amateur or professional, can work the land. Becoming a professional sylvanier has other advantages, too:

- Unlike farming, which requires daily monitoring most of the year, the sylvanier can take four months off in winter, and holiday time off throughout the year, without the forest garden suffering at all. This is something that will appeal to those outside the agriculture sector who are used to taking regular vacations.

- This activity is particularly suitable for those wishing to work part-time, keeping another profession or taking care of their families.
- A forest garden might fit in with most existing farm models and even benefit the areas dedicated to gardening by providing windbreaks, microclimates, biomass, and shelter for animals, as well as improving quality of life.
- Micro forest gardens can be implemented in existing parks and gardens, even in small areas.

Encouraging the proliferation of forest gardens could create many sylvanier jobs, and could help to stop the employment hemorrhage that has hit the agricultural world.

It should also be emphasized that this profession requires fewer technical skills than farming, as it is essentially low-tech. It does not even require extensive knowledge of ecology and agronomy, aside from knowledge of the complex tiers of the forest garden, which could be acquired in different ways (via short courses, guides, consultants, and technical advice). It is therefore easily accessible to people from outside the agricultural world, without requiring long studies.

The First European Forest Gardens

Among the forest garden resources available in England are two wonderful organizations: Martin Crawford's Agroforestry Research Trust and Ken Fern's Plants For a Future.[5] Crawford has imported hundreds of species and varieties of plants for fruit and berries from around the world, adapted to our latitudes and markets. Fern has identified and documented thousands of edible plants that could feed the world of tomorrow and created a resource center. In France, Franck Nathié has explored the concept of forest gardens for several years and offers training; he has also authored an interesting book on the subject.[6]

In Belgium, Gilbert and Josine Cardon planted a forest garden for the Les Fraternités Ouvrières (a brotherhood of workers) forty years ago. It is an amazing achievement on 2,000 square meters, completely outside the norm because of the density of the plants. Gilbert, a former mechanic, and Josine followed their intuition to create their edible forest in an exceptional spirit of giving — one based in the desire to make healthy food available to all. Gilbert, a charismatic man, says in the film *La Jungle étroite*: "I'd rather eat shit together than good food alone."[7] It's amazing what this fraternal forest produces: good for body and soul. Robert Hart, the food forest pioneer from England, was also motivated by a deeply humanistic vision.[8] He was interested in natural food and organic agriculture to solve the health problems that he and his family suffered. His research led him to study tropical forest gardens, and to adapt this concept to European latitudes. He implemented the first European forest garden on his small farm. His dream was to re-create the great original forest, to promote the planting of millions of mini forests, in town and in the country, to help restore the balance between the people and the land.

"The forest garden is the most productive use of land there is," noted Hart in his book.[9] "Most average about half a hectare (1.25 acres) in extent, and this small area can support a family of up to ten people.[10] They therefore offer the most constructive response to the population explosion. Java, home to the largest concentration of forest gardens — or pekarangan — is one of the most densely populated areas on earth. Yet the landscape does not have an urban feel because most village structures are built of natural materials and are hidden behind a dense curtain of vegetation."

Those now interested in the forest garden, and they are increasingly numerous as permaculture gains in popularity, are all custodians of Robert Hart's vision. "The forest-garden," he noted, "is much more than a system to meet the material needs of humans. It's a way of life; it also nourishes our spiritual needs with its beauty and the vitality of wildlife living in it."

At La Ferme du Bec Hellouin, we are now conducting a scientific study on the forest garden, in partnership with other sites that share our passion for fruit trees. This study will incorporate ecological, technical, and economic aspects of this particular form of agroforestry. A solid foundation of reliable data will bolster the expansion of forest gardens and foster the emergence of the sylvanier occupation. Forest gardens can contribute powerfully to feeding humanity in the future, while helping to restore the balance of our threatened planet.

Toward a Productive Agroecosystem, Resilient and Durable

At our current stage of research, it seems that a particularly mature form of agriculture — more elegant, perhaps — would be an agro-forestry-pastoral system relying heavily on the forest garden. In our view, it would be planted in bands, with fruit hedges forming narrow forest gardens from 5 to 15 meters wide, as we have observed that trees and shrubs usually produce more when they have good access to light. In our latitudes the light is less intense than in tropical environments, and access to this resource deserves special attention. These forest garden strips would crisscross the land, defining clearings that could become areas for food crops, cereal crops, or livestock. Ponds would supply the water needed for crops and animals. Such a system of miniaturized forests and clearings could be implemented in a small territory, from 5,000 to 10,000 square meters (1.25–2.5 acres), offering vast potential.

The benefits of such a system would be numerous. It would promote synergy among all its elements — trees, crops, animals, and human beings — and should prove to be highly productive.

What a joy it would be to work in such a space, both wild and lovingly gardened, away from the noise of the world! One hectare would produce an abundant and varied food supply, and provide jobs to several people. I like to imagine, in one of the clearings, a small drying and processing unit that uses solar energy, a hand-built workshop where crafts are made from natural materials like wicker. I see children swimming in a pond or playing in a cabin while their parents care for the gardens and animals.

Bill Mollison has said that we can also harvest pleasure from our gardens. The forest garden is a wonderful space. The one we planted six years ago proliferates every year. This is one of the places of the farm that we prefer. Despite the small amount of time we spend maintaining it, our edible forest is generous and gives us virtually no other work besides picking the crop. In spring, when the grass has not yet grown, we let the horse and ponies graze freely in it. Sometimes we will go there in the evening after school, with the children. As we pass through, our baskets are filled with juicy plums, vermilion apples, giant blackberries, aronia berries, and yellow raspberries.

I am pleased that our peasant profession allows us to reconnect with ancient techniques developed by first peoples, such as forest gardens or mounded beds. These agricultural practices are both extremely natural and highly productive; they are a priceless gift offered to humanity today.

Currently all the work in our gardens is based on agroforestry. We are convinced that trees will save the planet. Each of us should try to plant trees during our life as much as possible, everywhere. We are issuing a call for the emergence of a civilization of the tree.

SIXTEEN

Agriculture of the Sun

The crisis consists precisely in the fact that the old is dying and the new cannot be born . . .

— *Antonio Gramsci*[1]

Never doubt that a small group of thoughtful committed citizens can change the world; indeed, it's the only thing that ever has.

— *Margaret Mead*[2]

Adopting a widespread permaculture approach to food crops requires a profound break with the paradigms that underlie Western agriculture, a disruption of the concepts underpinning such conventional techniques. Can we take advantage of this disruption? If every human being adopted the lifestyle of a French person, it would take three planets to meet our annual needs. It would take four if we all adopted the consumption levels of US citizens. Because a minority of the world's population is consuming resources faster than they can regenerate, the world's human population consumes one and a half times what the planet can provide.[3]

Obviously, we're going to have to rethink our way of inhabiting the earth from top to bottom. We have no choice: There is no planet B. It is therefore necessary to invent novel ways of life that will help us, as quickly as possible, reduce our ecological footprint threefold.[4] The only reason that the current consumption levels are possible is because the vast majority of people are living in poverty: their reduced environmental footprint "offsets," only in part, our excesses. The 85 richest people in the world hold a fortune equivalent to that of the 3.5 billion poorest.[5]

Only One Earth for All

Reduce our ecological footprint threefold? Honestly, no one really knows how to do this, and even the most advanced ecovillages do not seem to have reached this goal.[6] One thing is certain: The measures taken by governments only minutely lighten our impact on the biosphere. The stonewalling on climate illustrates this point. Everybody agrees that we are heading for a wall, but how many nations are willing to decrease the competitiveness of their enterprises, making real decisions to reduce greenhouse gas emissions? The great problem of our time is that too few people take measures at the scale needed to effectuate change, and it takes courage and determination for those who really want to change their lives to distance themselves from the dominant system. Must we wait for major crises to finally decide to change our lifestyles?

Agriculture, the Bedrock of Our Civilization, in the North and in the South

To diminish our ecological footprint, it is appropriate to question all areas of our existence, without excluding any. But we are convinced that what we eat and how our food is produced deserves special attention. Revisiting the way we produce our food may well be the most powerful lever of ecological transition available: This is the area that will move all others.

Still, we need to avoid overly generalized, ethnocentric discourse. Food is produced around the world using extremely varied methods; agriculture does not lend itself to a single analytical framework. Mechanized agriculture, so widespread among us, is an exception. Globally, it remains the prerogative of a minority, according to the United Nations Food and Agriculture Organization (FAO). Of the current 1.3 billion peasant farmers, nearly a billion work only by hand; 430 million use animal traction. In the end, farmers with access to a form of mechanization are a fraction of this population, since there are only twenty-seven million tractors in the world.

Contrary to popular belief, it is not gigantic combine harvesters and GM crops in the Midwest, Europe, or Brazil that feed humanity, but primarily work done by the bare hands of peasants, mostly women. Permaculture represents a breeding ground of opportunity for these farmers, who do not have access to sophisticated technologies. It is gaining ground in Africa, Asia, and South America. Nepalese farmers have a practical guide to permaculture, the *Farmers' Handbook*, intended to "improve the well-being of millions of small farmers in the world."[7] Our permaculture friends in Cuba wrote a guide, *Permacultura criolla*, intended for farmers in Latin America.[8]

Eating Oil

Our Western agricultural system, whether conventional or organic, is responsible for an important part of our environmental footprint (about 30 percent, according to various studies).[9] This system is doomed in the medium-term future by its dependence on oil. In 2060, all the gasoline produced in the world will not even be sufficient to power the vehicles of the United States![10] And 2060, at the time of this writing, is in forty-six years, that is to say tomorrow. But it takes time, lots of time to develop alternative farming styles. This vast project should be initiated as soon as possible, considering the following:

- Oil supplies are dwindling. According to projections, world production will be 85.9 million barrels per day in 2020 and will drop to 17.4 million barrels per day forty years later.[11]
- In 1930, we used one barrel of oil to extract one hundred. In 2013, one barrel does not produce more than eleven.[12]
- "Having forty-five years of reserves, does not mean forty-five years of peace," reported the *Paris Match*. "The age of cheap oil is over no matter what happens. The easiest exploited oilfields have been discovered and production will depend increasingly on unconventional oil, which is costly and complicated to extract. In ten years, investment in research has increased more than fourfold, without increasing output."[13]

This last remark is significant because it describes an emerging phenomenon in contemporary agriculture and fishing: To maintain a constant level of production in a system based on an abuse of natural resources, inputs must be injected in increasing quantities, until the resources are depleted and the system collapses. At a certain point, the operating costs of procuring the resource are such that the system ceases to be economically viable. Since 1945, in the United States, pesticide use has increased by 3,300 percent, but pest-related crop losses have not diminished.[14] North American agriculture must now invest $2.70 of petroleum-based inputs to produce $4 worth of crops.[15] You can clearly see the limits of this approach: It will not survive the coming higher oil prices.

At La Ferme du Bec Hellouin, we constantly have these issues in mind. They motivate our decisions, including our decision to ban mechanization to the extent possible.

An Extractive Economy

Industry, agriculture, and commerical fishing are trapped in this endless "extractive" logic: They draw on the resources of the planet until the point of exhaustion. To satisfy our needs and desires, farmers squander the organic material resources of arable land, fishermen empty the sea of fish stocks, industry exhausts the mineral deposits of the subsoil, and we all collectively drain the fossil fuel reserves. We take without returning anything. We live on the capital of the earth, which is rapidly eroding. Such a policy is suicidal. The future lies in an ecologically sound, circular economy that creates value at every stage of the trade cycle, does not waste anything, and replaces what it takes from the earth. This type of economy is inspired by nature and is based on living systems, because only they are capable of generating a natural increase in resources. We will have to relearn how to live almost exclusively on biological resources, as our ancestors did.

The Food Challenge

Let's go back to agriculture. In this area, as in others, exploring new tracks is not a fad but a necessity: The food challenge we will face in this century is colossal. The equation is actually very complex.

- The FAO estimates that we'll need to double world food production by 2050 to feed a rapidly growing world population.
- Oil production, as we mentioned, will be increasingly diminished, which will greatly increase the price of oil.

- Arable land is shrinking fast: 30 percent of it is already desertified.
- Global warming poses unknown production constraints on entire regions, in the medium and long term.

Agriculture and Energy

The issue of energy is at the heart of agricultural transition. An abundance of oil usurped the human hand and animal power in the fields. Industrial agriculture is, from an energy perspective, nonsense.

There are three ways to measure the effectiveness of an agricultural system:

- **Yield per worker.** From this point of view, industrial agriculture is the best: Thanks to thermal engines, one farmer can handle areas up to several hundred hectares.
- **Yield per unit area.** Microagriculture is unbeatable, because the harvest is proportional to the intensity of care provided.
- **Yield per calorie invested.** Judged in terms of energy efficiency, our industrial agricultural model is an aberration. On a mechanized farm, 2 calories of fossil energy are needed to produce 1 food calorie (this inefficiency is partly related to crops intended for animal feed). If we add the energy to transport, store, refrigerate, process, pack, and distribute food, then 10 or 12 calories of fossil energy are required for a single calorie to arrive on our dinner plate.[16] By comparison, a study conducted in the 1930s in China showed that, for 1 calorie invested, farmers harvested more than 40.[17]

No civilization can endure when it allows itself to waste 10 calories to produce 1. Our system has developed into a massive process of unstocking the oil resources that nature took hundreds of millions of years to form. We live in the heart of a fireworks display that can only be short-lived.

A growing number of experts are sounding the alarm bell. If there were a sudden shortage of oil — linked, for example, to a geopolitical event — France would have only a few days of food reserves and could quickly sink into crisis. This may seem surprising to many readers. But yes, the specter of famine could recur in our developed countries. These nations are far more vulnerable today than before due to the rise of industrial and globalized agriculture, a colossus with feet of clay. That agriculture is based on a highly capital-intensive and centralized model: 80 percent of arable land on the planet used in intensive mechanized agriculture is owned by multinational corporations.[18]

These are strong reasons to reconsider the power of illusion engendered by the steel monsters that roam our desertified countryside. They are, in the image of the system that produced them, predators.

Inventing Post-Oil Agriculture

I have great respect for oil: it is a fantastic gift of nature. Oil is almost the energy of our dreams. It stores and transports easily, has countless applications, is simple to use, and creates breathtaking power. Yes, oil is a noble product that deserves our admiration. Knowing that we have only limited and nonrenewable stocks, we should use it frugally, reserving it for our most valuable needs, while making sure to keep as large of a stock as possible for future generations. Its power alone should make us cautious; we should not play with fire.

Yet we have chosen to squander this resource on just a few generations: ours, three before us, and — perhaps? — three after. Just yesterday, in historical terms, its use was anecdotal; our country functioned virtually without oil. Metropolitan France had only 1,672 cars in 1900 for forty million inhabitants![19] On September 2, 1914, a month after France declared war on Germany, the French army had 173 motor vehicles for 3.8 million mobilized men.[20] Our (great-) grandparents lived in that time not so long ago. If we step back and consider the recent history of humanity, the unstocking of the planet's oil resources is only a flash centered around the year 2000. Four million years of human and prehuman history passed without us or our ancestors using a drop of oil. It will now be necessary to collectively invent the rest of the adventure: the post-oil era — which will not be an easy task because we have developed, like drug addicts, a real addiction to black gold. Nothing will be like before. The work done by David Holmgren, the co-originator of permaculture, on peak oil and the energy descent of our civilization is of great interest.[21]

What will we eat in the year 2060? How will we heat our houses? What fuel will we cook with? How will we get to work, to school, go on vacation? What energy will make our clothes, our computers, and our consumer products? How will we receive our Internet connections, which represent a growing share of our ecological footprint? We have to find elegant solutions sector by sector. I welcome that. Long live the end of the oil era that has devastated the planet! I hope, however, that we will have the wisdom to retain the last barrels to facilitate the transition, create alternatives, and develop edible landscapes.

The energy descent that we are approaching will inexorably lead to a sharp reduction of the widespread transportation of goods and people, which is the norm today. In other words, we'll see the end of the globalization of material goods, including food. The end of the oil that allows us to feed our cows with GM soy from Brazil and export our wheat to Senegal. The end of the invisible barrels of fuel that allow shoppers to fill their carts, unashamedly, with lamb from New Zealand, rice from Thailand, Kenyan beans, Madagascan shrimp, Peruvian mangoes, Moroccan oranges, or bananas from Guadeloupe.

Relocalize Agriculture

The end of oil will have a major consequence: We will be required to relocalize our agriculture. Each region, each municipality must imagine a policy of food sovereignty and produce most of its own food (and energy, construction materials, fibers, and more). The ease and flexibility with which small farms can create microagriculture — and the low cost at which they can do this — will probably be the major key to creating local food security policies. The map of rural France in 2060 will hardly resemble that of today. The regions will host a dense network of tiny farms with diversified production in the extreme. Microagriculture will penetrate to the heart of urban spaces.

This change will have many consequences, including:

- Large farms will probably no longer be viable. They will be handicapped by their dependence on oil. The large volumes they produce will be too much for local markets and will require long-distance transport, a network of storage facilities, processing, packaging, marketing, distribution — all highly energy-intensive.

- In the future, small production and processing units, selling as locally as possible, will be the major option.
- To manage the microfarms of tomorrow, the number of farmers will have to increase significantly.
- We will probably see a partial de-professionalization of agriculture as individuals produce more and more of their own food, at home. This trend has already begun in the last several years in countries that have seen their energy resources diminish.

Each of these measures will contribute, for its part, to reducing our environmental footprint. We will move from a system that spends 10 calories to produce 1 calorie of food to production methods that will create tens of dietary calories for each calorie invested, like the Chinese farmers of the 1930s.

Toward an Agriculture of the Sun

Switching from an agriculture based on fossil fuels to one based on solar energy is not a panacea. Remember that nature generates a biomass per hectare larger than our best-performing cultivation practice without any input and has done so for hundreds of millions of years.[22] By mimicking nature, a permaculture approach to agriculture can allow us to create a world of plenty. But this will not happen without thoroughly changing the organization of our society. It is only at this price that we can move from an unsustainable civilization to one in sync with the biosphere.

Certainly at La Ferme du Bec Hellouin, our practices are still in their infancy, our experiments put together with bits of string. But we believe, like Victor Hugo, that the utopias of today are tomorrow's reality. In ten years, in twenty years we will have progressed, especially as exchanges with other explorers of the future are intensifying.

Completely rethinking our lifestyles in the terms of respect and balance is a wonderful challenge, drawing on all our creative resources. It is the most gigantic project in the history of humankind. It is something that can give hope to so many young people in search of meaning, desiring answers for what to do with their beautiful energy.

Choose Transition

But where to find the courage necessary for such a change? How to get out of the current stagnation, this lack of perspective and imagination? Protesting about future famine and environmental disasters is not necessarily the best approach. Perhaps it is better to offer solutions that generate enthusiasm, alternatives that make people want to make changes?[23]

No one is ever motivated to change if it means going in the direction of decreased quality of life — not farmers, not anyone. Try asking farmers in Beauce to exchange their tractor for a pair of horses! One of the main obstacles to the transition is the perception we have of it. We usually associate adapting to the limits of the planet with additional stress and deprivation, particularly in agriculture. One goes from "more" to "less."

Yet renaturing agriculture goes in the direction of less stress. The more we tame and protect our crops, the more they depend on us for their survival — leading to what Wes Jackson, founder of the Land Institute in the United States, christened the "treadmill of vigilance."

Change is not a punishment, it can be a liberating and exhilarating process. The ecological transition will improve a mediocre civilization that is destroying the only known living planet. We will have to deploy our potential as human beings to invent new ways of life, reconciling personal growth and the well-being of all. And we'll seek out a connected existence, rich with truly important things.

Do not wait for political and economic leaders to take meaningful measures. It would be great if they did, but they are often trapped in very short-term policy. It was under such a policy that Colombian authorities destroyed, in two years, more than 4,000 tons of organic seeds self-produced by small farmers — all because of its free-trade treaty with the United States, ratified in October 2011 by the US Congress, which requires seeds to be certified and registered in the official catalog, thus essentially produced by multinational corporations![24]

It is today that we must actively enter a transitional phase; 2030 will be too late!

The Path to Transition

Entering into transition does not mean resistance, a struggle against the old models. It means inventing the world of tomorrow, putting some color into life, moving toward something better. We do not imagine for a second that agriculture powered by the sun is a regression. On the contrary! It will be supreme elegance. Like the child Pupoli in his canoe, lightweight and efficient.

We have many strengths to succeed in this transformation. Our time is full of opportunities. In the West, citizens have never enjoyed such personal freedom; never have exchanges of ideas and positive solutions embraced the entire planet to this degree. We are in our best position yet to achieve an unprecedented qualitative leap. Moreover, the history of evolution shows that periods of relative stagnation — sometimes lasting for millions of years — alternate with periods of intense mutations, often related to crisis situations. The giant crisis looming is an opportunity. Ours is a very special moment in history, positioned between an old world that is not yet dead and a new world that is not yet born. Resisting this reality serves no purpose. Each of us can align our lives with our most precious aspirations. We can, in good conscience, take responsibility for our existence and the consequences of our passage on earth. Every person firmly committed to transition advances the entire human community.

SEVENTEEN

Working by Hand

Traditional agriculture was labor intensive, industrial agriculture is energy intensive, and permaculture-designed systems are information and design intensive.

— *David Holmgren*[1]

et's now turn our attention to how we can imagine an efficient and profitable form of agriculture based on the free resources of nature. Working by hand or with animal traction uses the energy of the sun stored in the plants and the animals that we nourish. This kind of solar power, on the scale of a human life, is infinite — constantly renewable — and its use does not cause any injury to the planet.

But gathering all our energy of the sun nevertheless poses a problem: It is diffuse, while fossil fuels — essentially solar energy concentrated over millions of years — are extraordinarily powerful.[2] Is it possible to change, in agriculture, from the use of a highly concentrated energy that is easy to store, to one that is diffuse and difficult to store, especially when tilling the land requires so much power? The answer, frankly, is *not easily*. But it soon becomes apparent that, in this context, to eliminate tilling is a huge step forward.

The Benefits of No-Till

To get a better idea of the relevance of work done entirely by hand, in our time, and evaluate its chances of economic success, it's worth taking a moment to reflect on the profound differences between no-till farming and mechanized agriculture. No-till is a practice inspired by the observation of nature: In wild spaces, machines don't plow the earth, yet it is effectively aerated and fertilized by countless life-forms, cost-free. In nature, the organic matter in soils increases slowly but surely until it reaches a state of equilibrium or combusts. It, too, is a concentrated form of solar energy, stored permanently in the humus. This natural increase in fertility is achieved without any external input.

Toward Self-Fertility

In a forest, as we mentioned, the biomass produced is generally twice that of cultivated areas without human intervention or inputs. This observation can be a powerful inspiration for farmers. Can we imagine a fully self-fertile agrarian system, one that operates like forests? This question is fascinating. However, there is a profound difference between our agricultural systems and natural ecosystems: Part of the biomass produced by the farmer is exported; it is not restored to the earth, as is the case in an ecosystem.

The question can be rephrased: In our cultivated plots, is the natural increase of fertility enough to compensate for the loss of fertility that arises from harvesting and distributing crops? The answer is complex. We must accept nature's guiding principles and observe the consequences of their implementation on the ground.

A review of the facts can inform our thinking. About 98 percent of an agricultural crop comprises items found in abundance in nature: water, nitrogen, carbon. The remaining 2 percent is minerals from the bedrock. The availability of all these elements depends on the intensity of the biological processes: The more alive the soil is, the more capacity it will have to transform environmental resources into plant nutrients.

The first objective of eco-farmers is therefore to foster soil that's as alive as possible. They can find bio-inspired solutions for avoiding the destruction of the existing fertility by:

- Using a no-till approach.
- Planting permanent ground cover, which prevents leaching and sterilization by ultraviolet rays.
- Using green fertilizers.
- Avoiding too much export of organic matter by restoring to the cropland all the nonedible parts of crops (leaves, stalks, straw).

- Leaving the roots in the ground, whenever possible, which among other things promotes mycorrhizal development.
- Favoring perennial rather than annual plants.

They can also increase the natural rate of humus creation by:

- Depositing decomposing mulch on the ground.
- Combining trees with crops.
- Raising organic crops.
- Composting in place rather than in a compost pile (a compost heap leaches nutrients and greenhouse gases, especially as it heats up).

Permanent mounded beds (adapted for vegetable and grain cultivation and certain forms of arboriculture) exclude the use of an engine, because a mechanical device, working the soil, would run completely counter to their purpose — namely, to develop the natural fertility of the soil.

The Engine, Soil's Worst Enemy

Any mechanical device passing through humus will undermine its fertility in the medium and long term. Statements like this may anger many farmers, as the tractor has become a symbol of the profession, but the following points are enough to persuade even the most ardent critics. *First, humus, the living and fertile component of soil, is created by plants, animals, microorganisms, and all the biological processes at work in the soil.* Earthworms play a key role in soil ecology. On average, the 250,000 present on 1 hectare (2.5 acres) ingest between 300 and 600 tons of soil each year.[3] In fifty years, a plot's entire mass of soil passes through the digestive tracts of earthworms, which led Aristotle to say, some twenty-four hundred years ago, that they are the intestines of the earth. There are (by weight) more earthworms in France than people on earth — an average of more than a ton per hectare.[4] At certain permaculture sites, like the Mouscron forest garden where the soil has not been worked for forty years, the weight of earthworms has reached 3 kilograms per cubic meter (or 6.6 pounds per 10.8 square feet)![5] But worms are vulnerable to the use of mechanicanized tools, especially rotary blades and tillers.

Second, the fertility of the soil is the fruit of the synthesis of its mineral and organic elements, a synthesis conducted by biological processes. It is estimated that at least 95 percent of arable land on the planet was created in forest-type environments.[6]

Third, soil is destroyed by machines. The repeated passage of mechanized equipment to prepare land for growing destroys the structure of the soil at the expense of its living components. Continuously working the soil with machines introduces significant amounts of oxygen, which accelerates the combustion of organic material. The carbon is released into the atmosphere as carbon dioxide, contributing to global warming. When the soil becomes deficient in organic matter, it is more vulnerable to leaching and erosion. The rise of industrial agriculture has contributed greatly to destroying arable land on the planet. Consequently, 5 to 10 million hectares (12.4–24.7 million acres) of agricultural land are lost every year in the world due to severe environmental degradation (excluding artificial soil).[7]

Why Plow?

Why, then, for centuries, even millennia, have farmers struggled to plow, to dig, to work their land? The answer is because they reap a short-term benefit, of course. Cultivated arable land tends to become compacted. As pioneer plants, weeds have a mission to colonize bare land. By removing all vegetation cover, the farmer creates favorable conditions for their proliferation. Plowing and digging allows the farmer, in a relatively simple, effective operation, to loosen the soil and remove the weeds. Moreover, the massive inflow of oxygen accelerates the work of bacteria and therefore the mineralization of organic matter — that is, it puts nutrients that the organic matter contains at the disposal of plants. The effect of tilling, therefore, benefits crop growth — as a first step. But in the medium and long term it works against soil fertility. According to Lydia and Claude Bourguignon, 2 billion hectares (4.9 billion acres) of arable land were lost in four thousand years of tilling.[8]

Plowing deprives farmers of the natural tendency of soils to increase fertility. They must then offset the loss of fertility: The system is under constant infusion. If farmers manage to find a balance in which the inputs offset the losses, their agrarian system can continue. This balance was achieved at certain times and in certain contexts. On the plains of northern

France and England during the second half of the nineteenth century, for instance, crops were alternated and pastures were well managed. But this is no longer the case today — thanks to the convergence of today's far more powerful machines and chemical fertilizers (which also accelerate the combustion of organic matter), and the resulting erosion of arable land.

Create Humus

Planting in permanent beds, with respect for the life of the soil, of course presupposes the abandonment of plowing and digging. Imitating, as much as possible, natural processes, the ecofarmer seeks to reconcile two seemingly contradictory needs: to get a bountiful harvest and to create humus to improve soil fertility. How is this possible ?

In nature, the pace of humus creation is generally very slow.[9] Good practices, such as those outlined throughout these pages, can accelerate this natural tendency toward greater fertility, a big step in the right direction. Additionally, the farmer can take further steps if the agroecosystem is not in equilibrium.[10] By turning farm waste into compost and using it as organic matter input, fertility is enhanced and value is added to the farm's resources. The creation of humus can be accelerated, relative to the slow rhythm of nature, by a factor of fifty to one hundred, depending on the amout of fertility transfer.

These fertility transfers deserve particular attention because they do not impoverish one part of the planet for the benefit of another — unlike North America's large-scale organic farms, which import most of their inputs. The sustainability and ethics of such a system are flawed. A permaculture farm, on the other hand, is focused on self-fertility, and so fertility transfers are more or less easy to accomplish, depending on the context. The necessary resources, if we are willing to look for them, are generally more numerous than we think. We find them in grass fairways, permanent green manure crops, tree leaves, hedge trimmings, ferns of the nearby wood. The spaces surrounding farms generally offer numerous free resources that largely compensate for the share of minerals that leave the farm when crops are exported.

We can also implement some of the fertility-building approaches that are more novel to our Western culture — like using effective microorganisms (EM) or biochar.

The Advantages of Small Size

Because permaculture allows us to produce a lot in a small space, that space is easier to fertilize than a large area. A fixed supply of compost goes farther. Spreading 10 tons of compost over 1,000 square meters (0.25 acre), for example, is ten times more efficient than spreading the same amount over 1 hectare (2.5 acres). This is one reason among many why it is possible to obtain exceptional soil with microagriculture. But microagriculture also presents another advantage: Keeping the cultivated area small frees up agricultural land for other uses — and especially for the production of biomass that can benefit the cultivated area. This makes it much more feasible for small farms to achieve self-fertility. The work of John Jeavons perfectly illustrates this point, and we will return to it in the following chapters.

Billions of Workers in the Black!

You might think, at first glance, that the balance of power between a mechanized farmer with a tiller or tractor and a market gardener working by hand is completely disproportionate. But you're forgetting that the market gardener has billions of invisible assistants: an almost infinite number of bacteria, worms, fungi, crustaceans, which help by secretly working day and night, including Saturdays, Sundays, and holidays, without ever asking for a salary or vacation pay, requiring no sick time or administrative work. How can it make sense to destroy such valuable helpers, then have to compensate for their absence by buying inputs?

The permaculture market gardener doesn't so much seek to grow plants. Plants know very well how to grow without a gardener; they have done so since the dawn of time. Rather, the gardener's efforts are focused on encouraging the development of all the life forces present in the garden. I like to think that we are the servants of earthworms! Agronomists call this taking advantage of ecosystem services — services that are, by definition, free and sustainable.

For all these reasons, agricultural practices based on the no-till approach can be efficiently executed entirely by hand on tiny surfaces and achieve a level of productivity that the machine cannot match. It is therefore possible to transition from an expensive and unsustainable method that is dependent on the combustion engine, to a clean, free, and sustainable sun-based approach.

The Relationship with Time

Choosing to base an agricultural strategy on ecosystem services rather than fossil fuels means accepting necessary changes to the scale of time. It means renouncing the brutal and immediate effectiveness of the internal combustion engine to enter a relationship with time modeled on nature's rhythms and cycles of life. Building a vibrant and diverse agroecosystem is not done by snapping fingers: it takes years for trees to spread their branches in the sky, for a pond to become fertile with diverse forms of life, for the soil to achieve the smell and texture of decomposing undergrowth.

Initially, the workload will probably be heavier and the return on investment slower in coming, although this remains to be seen. When creating a farm, the market gardener does not buy a tractor, does not build a shed to house it, but invests in hedges, fruit trees, ponds. Without big-ticket start-up expenses, market gardeners can focus initial efforts on creating raised beds. This can be achieved in a few weeks or months, with nothing more than a wheelbarrow and a shovel. Establishing the live elements of an agroecosystem, like hedges, trees, ponds, and water systems, is more expensive, but the investment in time and money can be made over the first five years, especially in winter, when crops require less work. In the medium and long term, the agroecosystem becomes increasingly self-fertile, resilient, and productive, and we can take advantage of the efforts made in the early years. Aging market gardeners can thus reap the fruits of the oasis of life that they have patiently created!

A professional approach should integrate the economic dimensions at every stage to ensure the sustainability of the company — something that Perrine and I had neglected to do in the early years. The main "trap" permaculturists can fall into is emotional: Their dreams are so fixated on the vision of a microfarm that the risk of embarking on the adventure without any technical preparation, or realistic estimation of the workload and stress, is high. We know one thing . . . the dream of returning to the land can become a nightmare. This is not to say that we should stop dreaming! Just be pragmatic dreamers.

At the risk of being a killjoy, I will also note that occasionally we encounter people who are animated by a beautiful ideal, but the gap between their aspirations and the contemporary world is so wide that I wonder if they aren't setting themselves up for failure. Some have created vegetable businesses while refusing to consider selling their produce: They want to live by barter. I cannot help but ask them if they can pay for their car, their rent, or their childrens' school supplies in salads and radishes.

One of the questions that motivated Perrine and I to demonstrate the economic relevance of a more ecological approach was this: Is working entirely by hand, in our profession of diversified market gardening and arboriculture, economically viable in the short term? Or will take it take ten or twenty years for higher oil prices to make it viable? Initial data from our study shows respectable results today for a method that yields high productivity, with low investment and operational costs, compared with a mechanized approach. But do not overlook the skill level required to achieve this result. You may need few tools, but you will require a lot of knowledge. Practicing agriculture by observing and imitating nature becomes a form of farming intelligence (fortunately accessible even to Neanderthals, I can testify!).

Hopefully, given the services rendered by microfarms, society will one day switch the allegiance now allocated to industrial agriculture over to new forms of farming.[11] This may take the form, for example, of financial assistance — providing initial investment support and basic remuneration for ecofarmers for their first three years, the time needed to create their farm and acquire the necessary expertise. Ecofarmers could also benefit from tax exemptions. When starting from scratch, becoming a market farmer is a risky adventure that deserves the support of the community — a community that will in turn be fed healthy and beautiful food, while seeing its environment enhanced. We are still far from this blessed time, but the success of la Foncière Terre de Liens, a community-based effort that buys land and rents it to new organic farmers, shows that awareness is emerging.

We are often asked how much capital it takes to create a microfarm. This number depends on so many parameters (size and locality of the farm, presence of structures or not, to name a few) that it is almost impossible to answer. It is easier to specify, for a market gardening activity such as that practiced at

La Ferme du Bec Hellouin, the investment in equipment and operating expenses. The ongoing study on our farm — analyzing the potential income of a 1,000-square-meter microfarm modeled after La Ferme du Bec Hellouin — should answer this question. It is always risky to give interim results, before the end of a study, but the relevant figures to date, although approximate, may be found on the farm website.[12] All the equipment and tools used in 2012 for vegetable production were determined to cost 22,000 euros (around $25,000, greenhouses included), for instance. This is significantly lower than the investment in equipment required for a mechanized installation.

The cost of creating a vegetable microfarm has so many variables that a study assessing just the financial risk involved in creating one would be very useful. At a minimum, buying a half hectare (1.2 acres) of arable land might cost 5,000 euros ($5,650); hedges, fences, pond, 10,000 euros ($11,300); a small storage building, 15,000 euros ($16,950); fruit trees and berries, 5,000 euros; equipment, tools, greenhouses, irrigation equipment, 25,000 euros ($28,250); used vehicle, 5,000 euros; miscellaneous, 10,000 euros; for a total of 75,000 euros (about $84,750). With this initial investment the farm can operate and be improved in the following years. Note that what is saved by the tractor and machine tool costs inherent in a conventional organic market garden approach is invested in trees, hedges, and pond construction.

These issues are complex, as one theme leads to another, but we will develop a practical manual covering the operation of a farm like ours at the end of the study.

A Change of Scale

At this stage of the book, you are certainly asking an important question: If microagriculture proves efficient for vegetable crops, how could we adapt the principles to large-scale crops, grain in particular, which require larger cultivated areas and can be difficult to grow without mechanization? We are not, at present, able to give a simple answer to this question, which remains largely outside our field of expertise. This point is not trivial, because it is an important challenge to the farming community. How will it conduct its grain business when fuels are scarce and expensive? How will growers harvest hay? Will we see the disappearance of tractors and combines, or even the end of large-scale farms, as predicted by Patrick Whitefield in *The Earth Care Manual?*

Producing biofuel on farms is one possible answer, but it does not solve the problem of obtaining the land and energy required to produce these biofuels, nor does it address the impact of biofuel production on soil, water, or climate. As they are currently produced, biofuels seem more like an attempt to perpetuate a flawed system than a real alternative. But things can change. Tomorrow's solutions will come from the soil, but also the oceans, which are a breeding ground for potential discoveries. Little explored to date, marine resources will certainly play a major role in the world to come. When society makes the choice to invest significantly in renewable energy, tremendous progress is made.

At La Ferme du Bec Hellouin, we welcome more and more people with large or very large growing areas, who are aware of the issues and willing to join us in networking with experts in regenerative agriculture from different continents. Around the world, many farmers and researchers are exploring alternatives: agroforestry; no-till, simplified cultivation techniques; cover cropping; and new ways of grazing cattle inspired by the great wild herds.[13] One of the most innovative lines of study is in the great plains of the American West, where agronomists at The Land Institute are trying to re-create perennial grains (the perennial character of the crop having generally been lost during its companionship with farmers) and perennial plant communities, including legumes, eliminating the need to work and fertilize the soil. This bio-inspired research is at the forefront of innovation.[14]

Here are some ideas percolating among those who are concerned about the future of agriculture. First, though, it makes sense to question not just the techniques of modern conventional agriculture, but also its outcomes. Does it really even produce healthy and abundant food for all?

Eating Differently

Our modern diet is based, much more than in the past, on a high consumption of animal products — particularly meat and dairy products. It turns out that the factory farming of animals is extremely polluting, a strong contributor to the negative environmental impacts of agriculture. It takes up to 10 plant calories to produce 1 animal calorie (this is the normal conversion rate in all food webs).[15] In fact, much of the grain produced worldwide is used to feed livestock destined for rich countries.[16] Moreover, excessive consumption of animal products is proving harmful to human health. The Good Planet foundation reported, "Sunday, February 2, 2014 was the day of the Super Bowl, the final game of the American Football League season. During this single day American viewers will have eaten 1.3 billion chicken wings. This number illustrates how our society and our daily lives rely on the mass killing of animals."[17]

It is clear that excessive consumption of processed grains and dairy products has been causing more allergies. Each of us knows children who are allergic to gluten or lactose, or even both, making daily life a nightmare as the use of these products is widespread in industrial food. The explosion of allergies and lifestyle-related diseases (obesity affects 36 percent of the population in the United States, a bastion of industrialized farming) shows that "modern" food, closely linked to industrial agriculture, is proving increasingly detrimental for the planet as well as consumers.[18] These are well-documented issues.

Conversely, research on the healthiest and longest-lived people on the plane — like the inhabitants of the valley of Vilcabamba in Ecuador, the Hunza Kashmir, the Abkhazians of the Caucasus, the people of Okinawa Island in Japan, and the Cretans — highlights what all these human communities have in common, which is, of course, living close to nature. Their diet is healthy, free of pesticides and herbicides, composed mainly of fruits, vegetables, legumes, whole grains, healthy oils such as olive oil and rapeseed oil, and pure water; consumption of meat and dairy products is limited or nonexistent.[19] Among these people who consider food as a "living medicine," the number of centenarians is particularly high; chronic illnesses that are constantly increasing in our Western society, such as cardiovascular and cerebrovascular diseases, cancers, obesity, osteoporosis, arthritis, and diabetes, are rare.

Therefore, before we think about how to change agriculture, isn't it necessary to consider another way to eat? Shouldn't we gradually revert to the food that sustained us during our long evolution, a diet to which our bodies are adapted: more fruits, nuts, berries, vegetables, leaves, legumes, and wild plants? By decreasing the share of meat and dairy products in our diet, we would improve our health and reduce grain requirements significantly. This effectively limits the amount of large-scale crop and livestock operations in our agricultural model, and therefore the use of fossil fuels.

Let me emphasize again this simple reality: What is good for us is generally good for the planet.

Toward a Permanent Agriculture

We are on the eve of a profound transformation of our agricultural model. The permaculture vision offers a strong route toward changing our practices, in a realistic and progressive manner, in the direction of sustainable agriculture. Here are some of the shifts we'll need to make.

FROM ANNUALS TO PERENNIALS

In nature, annual plants are an exception; perennials (with a life span of several years or more) make up to 99.9 percent of all plants.[20] Yet modern agriculture is highly dependent on annual plants, even though they demand much more energy and inputs than perennials. Paradoxically, too, 90 percent of the contemporary conventional agriculture is based on only twenty plants, especially grains: wheat, rice, and corn alone represent 60 percent of our food supply.[21] This depletion of plants grown in the modern era meets the profitability requirements of industrial agriculture at the expense of consumer health and agricultural systems.

It is estimated that there are thirty-five to seventy thousand edible plants; seven thousand have already been consumed by human societies.[22] The potential of "plants of the future" to diversify our agriculture is enormous. In this pool of edible plants, we will discover many perennials that require little water and fertility, with exceptional nutritional qualities. Consider the large seeds from the araucaria, or monkey puzzle tree, of Chile. Rich and delicious, they are traditionally consumed by the Mapuche tribes.

FROM STEPPE AGRICULTURE TO FOREST AGRICULTURE

Since its invention in the Neolithic period, steppe agriculture favors open spaces, plowed fields, and annual grains. An agriculture based more broadly on trees and the consumption of fruits, nuts, and berries would solve many environmental and health problems. The nutritional value of nuts, rich in minerals and omega-3 vegetable protein, is exceptional, and a hectare of chestnut trees produces as much vegetable protein as a hectare of wheat.

Wild plants generally have high concentrations of nutrients and strong health benefits, as demonstrated by a study of Cretan diets. They only require harvesting, no other effort, and do not leave the planet deteriorated. Introducing them, even in small amounts, to our diet makes sense.

It is therefore possible to imagine forms of agriculture that provide more space for trees and wild plants, are much easier on the planet, and better for human health, allowing us to move more easily away from the use of fossil fuels.

Among the many experiments under way in the world today, here are a few that can excite our imaginations.

FROM FACTORY-FARMED LIVESTOCK TO SUSTAINABLY FARMED LIVESTOCK

There are many animal husbandry innovations being adopted on small-scale farms across the globe. In England, Rebecca Hosking, a former television producer, took over the family farm and profoundly transformed its method of raising livestock, transitioning to a grass-fed operation. The farm's dairy cows no longer consume GM soy from Brazil, but remain in the field all year—a practice more common in England than in France, thanks to its rich and diverse mix of perennial grasses. They forage in the hedgerows that surround the meadows. The hedges diversify the cows' diets, improve their health, and even provide shelter. This form of farming, which requires very low inputs, approximates what was commonly practiced in our latitude before mechanization, and proves more cost-effective than intensive livestock practices. The revenues are lower, but investment and operating costs are as well. The breeder's quality of life is better, and opportunities to take a winter break are superior.

In Australia, Geoff Lawton, a globally recognized permaculturist, gives no more grain to his laying hens. They live in spacious movable cages, called chicken tractors. They nourish themselves with what they find in the soil, which they clear of pests and fertilize with their waste, as well as with compost, which is their main source of protein. Before the rise of industrial agriculture, hens were not fed grain grown specifically for them.

In France and elsewhere, extensive wildlife farms are sometimes established in forest areas. Deer, for example, can lead a natural and free existence before being shot by a guardian of the forest when he determines that the animal should be removed. This type of farming, when conducted correctly, does not degrade the environment. There is virtually no cost involved, and it provides high-quality meat. Even those who condemn the killing of an animal for food must recognize that the quality of a wild animal's life is infinitely better than our cattle experience in typical livestock operations. Raising cattle in a forest setting could meet some of the meat needs of those of us who are not, or not yet, vegetarians, especially if forests are becoming increasingly important in our farming landscapes.

In Corsica, Jacques Abbatucci raises his "tiger cows," a local breed with lovely tiger-striped coats, using the resources of the maquis, shrubbery that covers much of the island. His cows live outside all year and sleep under mastic trees, which protect them from flies. All of their food needs are supplied from within the farm. The animals graze freely under the canopy of the maquis, consuming wild plants that give their meat exceptional flavors. These wild plants supplement grasses grown in the alluvial valley of Taravo. The quality of Jacques Abbatucci's products is such that he cannot meet the demand.

Reducing our meat consumption could allow the elimination of cages, which are a disgrace to humanity, and allow the growth of natural forms of husbandry for the benefit of animals, farmers, and consumers.

FROM MACHINE-GROWN TO HAND-GROWN GRAINS

Since ancient times, grains were grown by hand, in gardens or small fields. According to Joseph Pousset, grain gardens reappeared in France during World War II.[23] It is generally believed that these practices are archaic and produce a very low yield. Yet research on wheat gardening showed yields comparable to the best yields of industrial agriculture. Imagine reducing our dependence on oil-dependent mechanization with a dual approach. First, by consuming less meat and ceasing to feed our livestock with grain, we would diminish over 60 percent of our grain needs. The grains we did produce would be reserved for human consumption. Then, by growing increasingly more grain on a domestic scale, as with fruit and vegetables, the need for mechanized crop production would be further reduced.

Furthermore, selecting and identifying perennial grain growing spontaneously in grasslands, such as wild triga, will lead to the emergence of a new type of self-fertile grassland, part of which will be wild perennial grains.[24]

Growing grain, milling it, and kneading and baking bread has deeply satisfying qualities. All over the world, there are horticulturalists, growers, and millers reviving forgotten varieties of wheat and other staple grains and processing them on a small scale. In southern France, in the village of Cucugnan, Roland Feuillas makes a "health bread" from ancient grains ground in a stone mill.[25] He has conducted extensive research to find ancient grains with outstanding organoleptic qualities, including local varieties that had disappeared, which he plucked from old thatched roofs. Feuillas cultivates his own no-till grain under layers of mulch. Customers flock to his exceptional breads.

We are convinced that an agricultural transition will not happen without a gradual evolution of the way we feed ourselves. Holding on to our very poor diet stands in the way of the many possible changes in agriculture. Again, it is up to each of us to change our practices, responsibly. Buying industrial food products perpetuates the industrial model. Choosing organic, local, fresh, seasonal products stimulates small-scale family agriculture, accelerating a necessary and enjoyable transition with many benefits at all levels, starting with our taste buds.[26] This healthy and delicious revolution depends on education and awareness; and this, frankly, is a big job. But the movement has begun and like Pupoli with his canoe, each of us must find our favorable vein in its current.

Collaboration with Top Chefs

We mentioned at the beginning of this book that it was our interest in healthy eating that led us to our organic vegetable business. There is an obvious complicity between the garden and the kitchen. Ecological farming can be generous and productive; it can also lend fruits and vegetables exceptional quality. From the beginning, and despite our inexperience, we received many encouraging words about the quality of our produce from members of our local agricultural association as well as shops and chefs.

In 2012 we began supplying some gourmet restaurants. The chefs came to us, and there were a fair number of them. Neither Perrine nor I had ever indulged in the gourmet food world: We had neither the means nor the desire, to tell the truth. Yet we unwittingly had created the base for a successful relationship with highly selective chefs. The thousand or so varieties of vegetables, herbs, fruits, and berries that we had come to grow, in pursuit of biodiversity, intrigued them. When leading chefs came to visit the farm, during visits organized by Slow Food for example, they were like children marveling at treasures. They told us that no organic farm had gathered in one place such diversity; it was their dream to choose among fifty herbs and almost as many edible flowers. "Your garden smells like a kitchen," stated one of them. However, many were not ready to abandon their usual wholesaler relationship in favor of a market gardener.

Stars in the Gardens

Our collaboration with chefs took a new turn during an encounter with Arnaud Daguin in December 2012. A mutual friend told us that a chef offered to cook us a meal at the farm, and one morning Arnaud arrived. We discovered later that he is a Michelin-starred and highly respected chef. With great simplicity and great humor, Arnaud set out to discover the gardens, a basket under his arm. In December, the garden is a bit sad and we were surprised to see him squat down, chew on stems and leaves, pull out his knife, and reveal the green meat of a kohlrabi or a radish. "Your vegetables are the bomb!" he cried out with his Basque country accent. The end of the morning was spent together in the kitchen. While preparing lunch, Arnaud told us about his passion for the world of small producers, which he knows particularly well. The meal that followed was, for the whole team, an unforgettable experience. Arnaud enhanced, in an astonishing way, these vegetables that we eat every day. Never before had they revealed such flavors! And yet his modes of preparation had a rare simplicity. Arnaud does not overpower his vegetables. His culinary genius places itself at the service of the produce, concentrating its flavors and mineral salts.

Two months later, Arnaud returned with a colleague, Antonin Bonnet, chef at Le Sergent Recruteur, a restaurant on the Île Saint-Louis in Paris, and the restaurant's owner. We spent the day in conversation. Antonin and Arnaud are talented chefs, but also have great knowledge of nutrition, able to talk about the molecular properties of each product and its impact on

The mandala garden is modeled on an age-old form. *Photography by Nicolas Vereecken.*

A group of trainees celebrates the end of their course by creating a small mandala garden. People from around the world come to the farm's onsite permaculture school to learn the Bec Hellouin method.

The base of this "hot bed" was constructed like those in Parisian market gardens in the mid-1800s. A pile of manure at its core keeps the soil at 68 degrees Fahrenheit (20 degrees Celsius).

The greenhouse provides a perfect environment for the chicken coop. No growing space is lost because there are raised beds above the coop.

Eliot Coleman visits a greenhouse at La Ferme du Bec Hellouin. An organic pioneer, he provided enormous inspiration to the Hervé-Gruyers.

From the fields, the abbey is visible above the trees.

The farm has become a community where family (*left*) and staff (*right*) share a mission to heal the earth while growing food.

The chickens leave their coop to roam among and fertilize crops by day.

The animals are sometimes as entertained by us as we are by them.

Constructing ponds was essential for improving soil and promoting biodiversity. Spiraling, raised beds allow plants to draw up nutrients from soils fed by dense pond life.

Every growing area is filled with food and flowers, including spaces around the farm's living areas.

Every growing area is filled with food and flowers, including spaces right around the family's thatched-roof home.

The family grew from four to six while the farm was in its early days.

The two island gardens are highly productive. *Photograph by Nicolas Vereecken.*

health. Both are also at the forefront of the new generation of farm-to-table chefs who believe chefs should adapt to the seasonal offerings of local producers, and not the reverse. Antonin's dream is to supply his restaurant from our gardens all year, preserving the surpluses of summer for winter. "That first day of exchange was, for me, a strange time, very special," recalls Perrine. "Without understanding all of what Antonin said about his cuisine, I had the impression that we were speaking the same language, him about his dishes and us about our crops. I pinched myself several times that day, hardly believing just how much we were on the same wavelength. When Antonin began speaking of lacto-fermentation, I thought for a moment that I was in paradise!" Several months and several visits later, we began an increasingly defined partnership with Le Sergent Recruteur. Of course we were a bit worried: Would our vegetables live up to the standards of a restaurant that the press was touting as one of the best? But after the first delivery, the owner sent us a text message: "I wanted to thank you because you have allowed us to move into the world of ultimate taste with your vegetables." We had never felt so respected as producers.

Antonin has invited us many times to Sergeant. Perrine and I are, without exaggeration, shocked by this experience. Our vegetables, our herbs, our leaves, all taste sublime. Sometimes we bring wild plants or flowers that are difficult to cook, like tansies. Even when our order is already under way, back they come, on our plates just half an hour later, perfectly prepared. A great art. We feel extremely proud of the mutual trust that has been established, and the relationship exemplifies a new type of partnership between restaurants and small organic producers — a trend toward harmoniously marrying a kitchen and a vegetable garden for the benefit of both parties. The volumes demanded by a single restaurant are significant and can provide a solid foundation for a market gardener, provided that the relationship is fortified with a contract, as few restaurateurs are truly ready to take into account the constraints of their partnership as Antonin does. This type of partnership is actually not that easy to implement. The chefs have to change their work habits, accept seasonality. The size of the produce is not always uniform. Weather conditions can cause setbacks. But they are rewarded with incomparable freshness and flavor. For us, as market gardeners, working with chefs drives us to constantly improve to meet their quality requirements. These expectations please us; they jive with our own quest.

We have not reduced the number of CSA baskets produced by the farm to accommodate them. On the contrary, the contents of our CSA baskets improve, too, from our work with the chefs.

But what matters most is that ecological farming can provide produce with peak organoleptic qualities. The chefs say that our vegetables have highly concentrated flavors and vital energy. We believe this is due to the natural conditions in which they grow. Our vegetables are not forced with an abundance of water and fertilizers, as is often the case, even in organic agriculture. They have to push their roots deep for water, and to reach soil that is alive with minerals and nutrients. Working with nature makes it possible to produce an abundance of good products, as evidenced by the growth of Slow Food around the world, which is a celebration of the talents of small farmers.

We are always curious to know how our colleagues are doing in France and abroad. New models are developed, tested, explored. Sometimes we encounter a particular technique that enriches the perception we have of agriculture and its integration in the contemporary world.

The Amish, a Fly in the Ointment of American Agriculture

In the United States, the jewel of industrial agriculture and production, there remains an interesting exception in several respects: the Amish. They come from a religious movement founded in 1525 in Switzerland. The first Amish disputed certain points of the newly launched Reformation by Martin Luther; they were violently persecuted by both Catholics and Protestants. Following this initial trauma, the Amish established themselves permanently on the margins of society and established strict rules to ensure the cohesion of their communities. To isolate themselves from the world, they all lived on farms and became excellent farmers. In 1760, the Amish immigrated to North America, where today there are over 150,000. Their lifestyle, in this early twenty-first century, remains close to that of the seventeenth: their clothing, language, and customs are the same. The Amish still resist modernity, materialism, and individualism.

Today, the Amish are almost all farmers and craftsmen, to the exclusion of other professions. They still live on farms, no longer small because they are home to several generations. They are milk producers and grow all kinds of crops, including grains, which is interesting because raising cattle and field crops is considered difficult without mechanization. But the Amish continue to reject what is now considered essential technology: mechanization, cars, electricity, telephones, television, Internet, although arrangements and compromises are sometimes adopted by communities.

"Today, the farm is the instrument par excellence of Amish survival and identity as a cultural group . . . All Amish see themselves as the manager of the garden belonging to God. As such, they must protect the earth. Overuse would be contrary to the divine will; the aim is to keep it in condition to meet the vital needs of man, but also to protect it against man. For the Amish, only work with the horse allows man to 'respect' the earth, the tractor 'crushes' the floor and does not respect the divine balance. Finally, the horse often outperforms the tractor, when climatic conditions make the ground too spongy," writes journalist Jacques Légeret, who made over thirty visits to Amish communities.[27]

How can an agriculture based on animal traction exist in the current context, and at a relatively large scale in some counties in the United States? The horse — or mule — actually defines the capacity of human labor. "What the horse can do in one day is sufficient for survival. The horse avoids excessive expansion of agriculture and the ongoing race to acquire the latest machines, with the debt that always accompanies it," observes Légeret.[28]

This type of old-school agriculture remains viable at the beginning of the third millennium. It even seems that Amish farms are generally more profitable than the industrial farms around them. As American farmer Gene Logsdon writes in *Amish Roots*, a collection edited by John A. Hostetler:

> The Amish have become a great embarrassment to American agriculture. Many "English" farmers, as the Amish call the rest of us, are in desperate financial straits these days and relatively few are making money. As a result, it is fashionable among writers, the clergy, politicians, farm machinery dealers, and troubled farm banks to depict the family farmer as a dying breed and to weep great globs of crocodile tears over the coming funeral. All of them seem to forget those small,

conservatively financed family farms that are doing quite well, thank you, of which the premium example is the Amish. Amish farmers are still making money in these hard times despite (or rather because of) their supposedly outmoded, horse-farming ways. If one of them does get into financial jeopardy, it is most often from listening to the promises of modern agri-business instead of traditional wisdom . . . More revealing, the Amish continue to farm profitably not only with innocent disregard for get-big-or-get-out modern technology, but without participation in direct government subsidies . . .[29]

In the country of big agriculture, where farmers represent only 1.2 percent of the working population, the Amish want to maintain an abundance of labor on their farms.

For the Amish community, the issue of employment is of great importance because it ensures the survival of its identity as a religious, cultural and social group. Indeed, the Amish have always believed that the father must stay in close contact with his children, as God wanted. Therefore nurseries and kindergartens are not allowed. It is a common sight to see little kids standing on the plows or corn carts beside their father. We remember a six-month-old infant who was blissfully taking a nap in a heap of corn stalks, gently rocked by the jolts of the cart and the horse's trot. For an Amish man, daily contact with his children is a sacred duty.[30]

The Amish model, based on strict communities, but warm and supportive ones, remains not only economically viable but also socially attractive despite its enormous stresses. Indeed, 80 percent of Amish young people today choose to perpetuate this way of life. The Amish reinforce our belief that other models — incorporating agriculture into community design — are possible. It's wise to remember that all great civilizations of mankind have been built on the most fertile land — and they collapsed when they exhausted it!

A New Era: Eco-Culture

Growing our food by hand is a utopia of sorts, a promise of a bright future. It is not a return to the past, even if post-oil agriculture assumes a form that resembles pre-industrial agriculture.

Since Neolithic times, farmers in Europe have almost always practiced various forms of tilling. The no-till, bio-inspired approach is very new in our country: It is a real technical and cultural revolution, which involves turning your back on a tradition practiced here for eighty-five hundred years.

The changes that will take place in the next fifty years will be so deep that we will probably refer to that time as the end of agriculture, and the beginning of eco-culture. Agriculture involves transforming natural ecosystems into artificial production areas that serve humans. Eco-culture proposes to take the natural ecosystems as models, and seeks appropriate responses to human needs with respect for other forms of life that are perceived as allies, favoring our survival.

By combining the best from tradition and the best from modernity, we will have great tools to invent the eco-culture of tomorrow. Each permaculture garden contributes to the reconstruction of the vital potential of our planet.

==== EIGHTEEN ====

To Be Small

Give an individual guaranteed possession of a barren rock, he will transform it into a garden; give him a garden with a nine-year lease, he will turn it into a desert.

— *Robert Blondin*[1]

*I*f, due to the coming energy descent and the efficiency of microagriculture, working by hand became widespread, it would inevitably lead to a significant reduction in the size of farms. We cannot carefully work large areas by hand. Being small presents many advantages. Most permaculture principles are applied more easily on a small scale. It is easier to increase biodiversity, create microclimates, and practice multiple activities on 1 hectare than it is on 100.

So it stands to reason that, in the developed world, tomorrow's agriculture will be a revival of small farms going against the current industrial trend of ever-larger farms. But it's worth remembering here that, on a global scale, the vast majority of farms are already microfarms; 90 percent are under 2 hectares (5 acres). These microfarms exist mainly in the Southern Hemisphere, where they are worked by a billion peasants without any form of mechanization.

Microfarms in the Southern Hemisphere

This opens the door to the potential of a bio-inspired approach to agriculture in Asia, Africa, South America, and Oceania. A neglected potential, too: In 2006, agriculture accounted for only 3.7 percent of all expenditure on public aid for development.[2] These billion bare-handed farmers could relatively easily adopt — if they haven't already — agroecological and permaculture practices.

Small farmers in poor countries usually lack capital to invest in seeds, or in machinery fom the West, but they have a very good vernacular knowledge of their environment; they have lived in it intimately for so many generations. This knowledge is a precious treasure, but unfortunately an untapped one, discarded by the trend in the Southern Hemisphere toward Western agricultural practices. This push toward industrialized agriculture, so far removed from traditional agriculture, leaves small farmers feeling, at best, in need of assistance, and at worst ignorant, dispossessed of their dignity. They could nevertheless rely on the power of their own labor, intelligence, and ancestral knowledge to develop agriculture that is bio-inspired and tailored to their region.

Permaculture farming is low-tech in terms of technology, but it requires a sharpened understanding of the natural world. It is an agriculture based in knowledge that can't be generalized, except in a decentralized manner that leverages the intelligence of local communities. Its development does not require large investments, only suitable education programs to develop the skills of traditional peasant communities and enrich contemporary knowledge of agroecology and permaculture.

Hungry Farmers

We are still a long way away, alas. Today, nearly one billion people suffer from hunger in the world. Paradoxically, 75 percent of the undernourished are farmers, food producers often ruined by unjust globalized trade and inappropriate policies.[3] When the wheat and chickens subsidized by the European Common Agricultural Policy arrive at the cheapest prices in the African markets, farmers in Senegal and elsewhere have little chance of surviving this unequal competition.

It is said that in northeastern Brazil, a land subject to periodic famine, and at the time of this writing undergoing its worst drought in thirty years, mothers cook stone soup. When their children cry from hunger, they reheat their pots and replace the water and stones. "When will the meal be ready?" ask the little ones. "Soon, soon," whisper the mothers, hoping that their children will fall asleep by dint of waiting. What must these parents feel as they attend, helplessly, to the decline of their children?

The families of farmers who cannot live with dignity from their earnings in the village are often forced into selling their land and often meet failure in the slums of megacities. Such was the fate of farmers in Maharashtra, India. All the major slums of the planet are born of the massive exodus of farmers from rural populations. There, where violence and lawlessness are frequent, families quickly go adrift: The father drinks, children live on the streets, sometimes forced into prostitution to survive, while mothers throw all their efforts into a hopeless battle to save their little world. What would it cost to reintegrate them? The components of such an effort should be carried out all at once: health programs, relocation, education, job creation. Yet only small sums spent on educational programs conducted upstream, in the heart of rural communities, would enable farmers to better understand how they can make a living from the surrounding environment, valuing the services provided by ecosystems. What suffering could be avoided! Examples of such achievements are not lacking in the world. The Songhai Centre in Benin, which offers training in sustainable agriculture — indeed in a sustainable system, incorporating energy and economics, too — is one of the most beautiful examples.

The industrial approach to agriculture, the so-called green revolution, is based on imported technology and creates dependency and vulnerability. The agroecological and permaculture approach is based on education and the enhancement of existing knowledge; it strengthens autonomy, food security, and the dignity of farmers. It also holds significant promise for increasing food security. The many studies of agricultural systems in various countries have, for decades, generally agreed on this point: The smaller a farm is, the more productive it is per unit area.

The World Census of Agriculture conducted by the FAO and other recent works by international institutions both confirm this. Farms between 0.5 and 6 hectares (1.2–15 acres) have proven to be on average four times more productive — and sometimes twelve times more productive — than farms with more than 15 hectares (37 acres).[4] In the United States in 2002, the revenue per unit area for farms with less than 1 hectare (2.5 acres) was ten to fifty times higher than the largest American farms.[5] This is often described as the "negative relationship between productivity and farm size."

Microfarms, in an Urban Setting

There is a wonderful thing in microagriculture, which is that potentially *millions of farms are sleeping everywhere, even in the heart of cities*. If a cultivated area of about 1,000 square meters (0.25 acre), as in La Ferme du Bec Hellouin, produces, throughout the year, the equivalent of sixty to eighty CSA baskets of fruits and vegetables weekly, then any garden, even a small courtyard, becomes a potential farm.[6] A 200-square-meter lawn, with good soil and a skilled gardener, could be producing a dozen weekly baskets. A 10-square-meter balcony, almost a basket. Flat roofs represent so many opportunities.

Let's imagine that the energy descent is accompanied by a sharp decrease in the number of cars and trucks. Tens of thousands of hectares of roads, highways, and peripherals would be liberated! Close your eyes for a moment and imagine the footprint of cars on your city today. Then imagine your city and its surroundings in twenty or thirty years, served by an efficient network of public transportation and bicycle paths. The arteries, squares, parking lots could be transformed into as many green corridors, vegetable gardens, and forest gardens. Songs of birds and the scent of flowers would replace the exhaust and visual aggression, the sounds and smells of vehicles. Then we would realize what violence the age of the combustion engine had exposed us to.

It is in the cities and their suburbs that the potential for microfarms is the greatest. By finding more elegant solutions than cars for our transport, without demolishing any buildings, tomorrow's cities could achieve genuine food resilience. Replacing part of the gray asphalt with green flowing networks that irrigate the urban fabric like so many arteries, veins, and capillaries would create a new type of city. A welcome slowness, like the one advocated by the Slow City movement. A foreshadowing of this change is visible in the US city of Detroit, which is attempting to survive the collapse of the auto industry by cultivating its wastelands. Each gap in the urban fabric is a potential garden. Tomorrow's farms are already there, before our eyes, in the making. A detoxification of the land may be required, and can be achieved, in many cases, with plants or pollution-cleaning mushrooms.

I do not believe at all in futuristic projects involving vertical farms, sophisticated systems, artificialized and energy-intensive. These will in no way meet the challenges of the upcoming, unprecedented energy descent. It is so much simpler and less expensive to carefully cultivate small spaces. In 1850, the population of New York (one million people) sourced its food from a radius of 10 kilometers (6 miles) around the town.[7]

Microfarms and Companies

Some companies are beginning to look at transforming the spaces surrounding their headquarters or factories into permaculture gardens. Potentially, thousands of hectares of lawns and wasteland could be easily converted into edible landscapes. Three market gardeners trained at La Ferme du Bec Hellouin are currently helping the Manutan company with a visionary project: surrounding their warehouses in Gonesse, France, with a garden whose production will supply its canteen. Employees can garden during their working time and benefit from educational support offered there. Such a project is likely to improve the quality of life and the social climate of a company, not to mention the positive impact on its image, both internally and externally. Employees are proud of their company when it is committed to the environment. Everyone wins in this type of adventure.

New Farmers

Producing food in the heart of cities goes far beyond the need to feed the population. Farming or gardening is a lifestyle, an aesthetic, a political and spiritual approach to life. The farmers of tomorrow will not be coming from the agricultural sector, which has been reduced to a trickle. The neo-farmer will perhaps be your office mate or your next-door neighbor. Your spouse or your child. Who knows? You can easily imagine that they will not spontaneously gravitate to the industrial model with its GMOs, pesticides, and sputtering machines. They will be drawn toward what feels good to them and to others. They will lovingly cultivate the land.

These farmers of tomorrow will probably practice a multitude of jobs: doctor, teacher, plumber, physiotherapist, or beautician three days a week, market gardener on the other days. Becoming a market gardener may also represent a great opportunity for the unemployed and for those over fifty who have scrapped the world of work.

All those who now have a suburban house or a small garden can make this transition happen overnight, with virtually no investment, and feed five, ten, fifteen surrounding families. What balance of life they will find in this transition! What a pool of job creation at the heart of cities!

Those who have no land can forge fruitful partnerships with local land groups, which are increasingly seeking to develop local organic food resilience. The management of some parks, wastelands, sections of pedestrian streets, and roofs could be delegated to individuals or associations, with a mission to turn them into food gardens to the delight of all. These spaces then become productive and would gain in attractiveness. They fulfill multiple functions in the service of the community. It turns out that it generally costs less to maintain common vegetable gardens than parks, and greening up cities is one of the most effective ways to reduce vandalism. In hospitals, it has been shown that patients with views of a garden rather than a gray wall see the length of their stay reduced by one day on average.

And in the Countryside

In the countryside, space is more readily available. The model of microfarms we imagine is not limited to 1,000 cultivated square meters — quite the contrary. Microfarms of 1, 2, or 3 hectares (2.5–7.5 acres) hold a surprisingly rich potential.

Without wishing to denigrate in any way the dominant organic market farming model of today, I would like to highlight the opportunities offered by a permaculture approach.

One hectare (about 10,000 square meters, or 2.5 acres), we have seen, is the minimum area considered necessary to install a mechanized organic vegetable farm. In this configuration, the gardener has to cover the entire surface, or nearly, with crops.[8] There is little room to plant trees, hedges, dig ponds, build a home. The farm then remains a highly artificialized area, even if it is organic.

Take the same hectare, remove the tractor, and focus vegetable crops on approximately one-tenth of its surface. Imagine all you can do with 9,000 square meters that have been freed up. Outline the plot of fruit hedges, plant a forest garden on the side of the prevailing winds, build a bioclimatic house for you and your family, dig ponds for an aquaculture practice (or production of spirulina if the weather permits), plant an orchard, keep a cow, sheep, or a draft animal, install a chicken coop and beehives. The list could go on. It becomes possible to create a truly diverse agroecosystem, providing work for two, three, or four people. Economic security is increased because of multiple activities. It becomes possible to create a small point of sale for this diversified production.

Bringing together crops, trees, and animals allows the microfarm to become independent in organic matter, by taking advantage of the multiple resources available. Such a place is also beautiful and provides a good quality of life.

The permaculture approach enables us to realize our complete vision of agriculture. To move from one paradigm to another, it is appropriate to dust off our imagination. Some find it a real joy to invent the alternatives of tomorrow; others experience a kind of terror at the thought of leaving the old models behind. In truth, a journey that takes us from the known to the unknown is always a risky adventure. But we cannot forgo this trip. Life is made up of death and rebirth cycles: It's the price of growth. Problems arise when we resist necessary changes. Civilizations do not escape this rule. Energy descent condemns our current model, so we are forced to migrate to a new future. Rather than clinging friviously to outdated concepts, choose to move forward, joyously, starting today. This is the essence of transition: Consider it a fantastic opportunity to be able to care for the earth and its people.

NINETEEN

Microfarms

I suggest that the foundations of peace cannot be laid by universal prosperity, in the modern sense, because such prosperity, if attainable at all is attainable only by cultivating such drives of human nature as greed and envy, which destroy intelligence, happiness, serenity, and thereby the peacefulness of man . . . The cultivation and expansion of needs is the antithesis of wisdom . . . Only by a reduction of needs can one promote a genuine reduction in those tensions which are the ultimate causes of strife and war.

— *E. F. Schumacher*[1]

A globally non-generalizable civilization is morally unacceptable.

— *René Dumont*[2]

I will never forget my meeting with René Dumont, the agronomist who was one of the pioneers of environmentalism and alter-globalization in France.[3] It was 1992. The old lion was then eighty-eight years old. I was preparing an expedition to Guinea-Bissau and he had agreed to preface our book.[4] This man who trekked all over the earth was having a hard time walking, but under his white mane, his eyes had lost none of their youthful ardor. During our exchange, René insisted forcefully that the Western agricultural model was flawed as much technically as environmentally; moreover, he said, it was morally indefensible, because it assumed a continuous flow of raw materials and foodstuffs from across the world to the 20 percent who consume 80 percent of global resources. Our wealth was constructed at the expense of peoples of the Southern Hemisphere. "These are the same mechanisms of domination," emphasized René, "that exploit people and plunder nature."

At the time of this meeting, René had not mentioned the oil dependency of our agriculture. Awareness of energy issues was not as prevalent back then among the French as it was among those in the Anglo-Saxon world.[5] What would he say today now that all environmental indicators are highly degraded, hunger has progressed even more, and the gap between rich and poor has grown? On average, a resident of the United States consumes eleven times more resources than a Bangladeshi and one hundred times more energy.[6] We are en route to a planetary crash of an unprecedented magnitude.[7]

The solution will not come only from above, from major programs subsidized with billions of euros. It will come from citizens, in a flowering of local and decentralized initiatives. In troubled times, times of war or economic crisis, when states are so frazzled that they can no longer feed their population, microagriculture and family gardens are always the last resort. This was true on many occasions. In the United States, for example, when Eleanor Roosevelt urged citizens to grow Victory Gardens during the Second World War; in Russia during the collapse of the USSR; in Cuba during the same collapse and while the US embargo brutally deprived the island of its imports; in Greece today. Microagriculture permaculture seems to be an elegant way out of the impasse in which we are embroiled. With sober means and maximum efficiency, it has everything to seduce the builders of the future to enter a virtuous spiral.

Many Possible Activities

In these pages, we have so far described microfarms focused on vegetable and fruit crops — operations like ours at La Ferme du Bec Hellouin. But many farming operations exist in areas of just 0.5 to 3 hectares (1.2–7.5 acres). What can a farm that size focus on? Here are some examples, and with a little imagination you can expand the list:

- **Arboriculture with on-farm processing.** One hectare (2.5 acres) of organic apple trees, for example, can produce 20 tons of average apples, which can be processed into apple juice, cider, sauce, jellies, jams, and compotes, or dried.
- **Berries and berry processing.** Small organic fruit production benefits from much research, and commands a good market price. Berries can be made into a wide variety of delicious products.

- **Small livestock.** Poultry, sheep, goats, pigs, and rabbits can be reared in a small area, possibly in combination. Eggs, milk, and meat can create added value.
- **Beekeeping.** Hives require only a small area.
- **Aromatic and medicinal plants, mushrooms, and flowers.** These crops are labor-intensive, but take up a small area.
- **Nursery, production plants, seed production.** As above.
- **Bakery.** A small area of cultivated grain is enough to supply the activity of a baker.
- **Aquaculture.** The potential productivity of an aquatic surface is several times higher than the same land surface.
- **Educational farm.** This type of farm produces less food, but fulfills an essential social function.
- **Farm bed-and-breakfast.** Events and ecotourism are better suited to small, well-cared-for spaces.

The potential is vast. Almost all of our food could be produced by hand in small areas. The secret to making a microfarm economically viable seems to be twofold: multiple integrated activities (chickens, bees, and berries can coexist in an orchard, for instance), and on-farm processing to create value locally rather than in factories. Only two types of production seem difficult to reconcile with microfarms — cattle breeding and grain crops. We'll explore alternatives in the pages ahead.

The Ecosystems of Microfarms

The microfarm is a small world in itself. If its potential is vast, it can be increased tenfold when all its components are integrated. The essence of permaculture is to connect elements so that they interact to form a complex system, like the broader surounding ecosystem in which the whole is more than the sum of its parts. Our intuition tells us that the microfarm is only the first stage of the rocket. The second stage would be the ecosystem of the microfarm.

Imagine a small territory in the range of 5 to 15 hectares (12–37 acres) on which farmers could join forces to launch a set of complementary activities. We could find a market gardener and an arborist, a baker and a beekeeper, a producer of herbs and a poultry breeder, a goat cheese maker. What benefits would they find in their association?

- In terms of ecosystem services, the act of involving a wide variety of activities in a small territory enriches the dynamics of the whole. The green waste from vegetables feeds the chickens and goats, whose droppings will in turn enrich the land. Bees from the beekeeper pollinate the aromatic plants, fruits, and vegetables. Fruit trees will block wind to help the crops, and so on.
- On the economic front, a number of functions and equipment could be shared: a draft animal, a sprayer for the liquid fertilizer, a cold room, a common point of sale for all of the farmers, an accounting clerk, a joint website. The cost of this investment is spread over several farms instead of weighing on one.
- In terms of quality of life, the association helps members support one another, achieve some major work in common, and take vacations in turn.[8]

Such a microfarm ecosystem must find appropriate modes of governance. It seems most desirable for everyone to manage their own activity with maximum autonomy, and possibly own their land along with its infrastructure and tools. The truly collectivist agricultural organizations often prove unproductive and seldom withstand the test of time. A collective farm project should have a long duration and must be able to overcome crises and departures.

The microfarm ecosystem will create sustainable, local jobs at a modest cost — far from what might happen if the land was managed by the canons of modern agriculture.

Agrarian Solidarity Systems

But we can go even farther. Imagine the third stage of the rocket. One day I explained this vision to Marc Dufumier, who replied: "That's what you have to do. That's called an agrarian solidarity system." The third stage was baptized.

I wish I could take a sheet of paper and draw my vision for you. The idea is very simple: A large number of complementary agricultural activities merge together on the same territory — ideally encompassing all that the region is capable of producing based on its climate. The figures in the hypothetical scenario that follows are merely indicative. There will need to be studies to confirm what is now based only on my intuition. Don't take the numbers below literally; just try to visualize the construction of the agrarian system they help define. You can enrich the vision with your own ideas.

Take an average farm of 100 hectares (247 acres); a round number facilitates our demonstration. It could be an existing grain farm, for example. The 100 hectares are covered with two, three, or four crops that give work to a single farmer. Now imagine that a strong team has decided to take this farm and transform it into a permaculture-based agrarian system: you, me, and thirty other people willing to roll up their sleeves; youths and seniors, unemployed people, and those who decided to leave their job to live closer to their ideals. All bring skills, talents, creativity. The farmer turns over his operation and we create a design together for our farm.

A constant principle will guide our work: Get the most out of the services rendered by ecosystems. We will therefore seek to renature the place by applying the principles of permaculture described in this book. Because many of us will take care of this land, we can plant it densely, so we make space to plant trees, in large numbers, in all possible forms. Our goal is to achieve a self-fertile system in the medium term. After a year of intense reflection during which we together drink a few good bottles, get our brains warmed up, and work with computers, our design is finally ready, beautifully formatted in a well-illustrated folder. Which is what convinces the local authorities and various partners to follow us in the adventure.

Our agrarian solidarity system prototype will be used to model many similar experiences. Scientists from various disciplines are closely involved in its design and in its implementation.

Cover All the Nutritional Needs of Local Communities

Early on in the process, we agree that the primary objective of our system is to produce an abundance of food, with excellent variety and quality, and to cover all the nutritional needs of the local community. Certainly we will not produce bananas or vanilla or coffee, but we will produce all that is necessary for a balanced and healthy diet.

We have yet to embark on the fieldwork, but our goal is to transform the dreary grain field into an edible landscape of astonishing productivity. Here is our plan.

A PRESERVE FOR BIODIVERSITY

Priority is given to the regeneration of nature: 5 hectares (12 acres) are devoted to the protection of biodiversity. This sanctuary is planted with local species, a pond is finished, and then we let nature evolve as it will. We will close it to human use — except for possible scientific pursuits.

AN EDIBLE FOREST

Forty hectares (99 acres) are planted with trees carefully selected for the services they can offer. This forest requires some development, since it is the most innovative part of our integral agrarian system.

An edible forest, as I see it, is different from what we generally mean by *forest* in Europe, where typical forests produce very little food (though during some troubled times, populations have permanently lived in the shelter of the woods, demonstrating that their potential is underutilized). Our current forests mainly produce wood. In other parts of the world, forests contribute much more to local economies. Tropical forests provide a wide range of products: wood, of course, but also fruits, nuts, various edible or medicinal plants, gums, fibers, building materials, and craft supplies. These tropical forests support the activity of people who collect these various products, such as the farmers of the Amazonian forest from Brazil, who are the equivalent of the sylvaniers we want to appear here. The edible forest is inspired by tropical forests, even if it consists of native plants for the most part.

The edible forest also differs from what we described under the term *forest garden* in chapter 15. In the world of permaculture, these terms are used interchangeably. We propose to introduce a semantic distinction that will more accurately describe systems that are very close in their essence but with a difference in scale:

- **The forest garden** (or food forest or garden forest), described in chapter 15, is a small agroecosystem, intensely neat, in which the canopy is formed mainly of fruit trees. Robert Hart also called it a mini forest, emphasizing the miniaturized aspect of this layered plant community, which rises to 10 meters tall at its full development.

- **The edible forest** is itself an evolution of the forest garden, but larger, from 1 hectare to several thousand. The tree layer consists tall-growing species, mainly trees bearing nuts. The intermediate layers are less productive. The edible forest is designed to offer a wide range of services to humans. It is a less intensely gardened agroecosystem than the forest garden, but neater than our customary forests that are hardly used (by us) beyond logging and hunting activities.

Here are the characteristics of the edible forest, as I now define it:

- The canopy is mainly composed of trees bearing edible nuts for humans and animals: walnuts, chestnuts, oaks. We have already mentioned the great nutritional value of nuts and the desirability of including more of them in our diet.
- An intermediate level is planted with shrubs for artisanal crafts or nutrition, such as hazelnut.
- A lower layer can accommodate (in clearings) bushy plants with berries (blueberries, currants, raspberries, blackcurrants, and many others from diverse regions of the world; remember that berries generally originate from undergrowth).
- The forest also produces a large number of wild edible plants. Mushrooms can be grown.
- It can serve as a breeding place for animals living in semi-freedom: pigs, deer, cattle, chickens, for instance.
- The forest provides firewood and lumber for eco-construction. Chestnut trees are well suited to the manufacture of shingles — "tiles" of wood giving long-lasting and beautiful shelter.
- It provides materials for handicrafts (baskets, fences). Cuttings that stimulate forest growth can provide young straight poles that have multiple uses.
- Some plants can be used for their fiber, such as brambles (used in baskets) or nettles, to produce extremely resistant fabrics.

- Part of the biomass can be enhanced by anaerobic digestion (gas and heat) or biogas, and used to fertilize croplands. Ferns can be used for mulch and animal bedding.

Thus considered, the edible forest plays a very important role in ecological balance and the economic and social solidarity of the agrarian system. It provides one or more jobs. These sylvaniers are not unhappy with their fate, for managing a forest is a vibrant and wonderful job.

Those who wish to further explore the potential of forests in a permaculture vision can read the excellent work of Ben Law.[9] In England, for a generation, we have seen a real revival of green-wood crafts. Green wood is soft and easy to work. Many common, strong, and beautiful objects can be made from green wood as an alternative to plastic industrial objects. In France, this revival is still in its earliest stirrings. This is an exciting avenue to explore, an old business model to revive.

So, between our 5-hectare forest preserve and our 40-hectare edible forest, we have reforested 45 hectares. Let's see what we could do with the rest.

A GRAIN FARM

Twenty hectares (49 acres) are dedicated to grain. Here are the elements that comprise our design for grain on this farm:

- Grain crops. In their conventional annual form, these crops are highly dependent on oil, so we would reduce their production as much as possible for now, and hope to grow perennial grains in the near future.
- We discussed the energy aberration involved in cultivating grain to feed cattle, which currently consume two-thirds of the grain produced worldwide.[10] Grain will therefore be exclusively reserved for human consumption, reducing the surface area dedicated to it.
- The emphasis given to nuts will allow us to combine grain flours with nut flours to make bread, again reducing grain needs. (Nut flour cannot be used alone for bread making.) Increasing use of chestnut flour will be beneficial to human health because of its nutritional power.

- This farm will also produce oil (rapeseed, sunflower, and flax oil, for instance). Again, the edible forest covers part of the oil needs through walnut oil, hazelnut oil, and high-quality almond oil, which will allow us to reduce the amount of oilseed crops.
- Until viable alternatives are found, grain will be grown in our permaculture system as virtuously as possible: using agroforestry, sowing under layers of mulch, employing simplified cultivation techniques and, if possible, animal traction.
- Agroforestry techniques are now well established.[11] The rows of high trees in the fields will preferably consist of walnut and chestnut, so as to substantially increase the production of nuts of our agrarian solidarity system.
- After the harvest, the fields can be grazed by a flock of sheep.

A CATTLE FARM

Twenty hectares (49 acres) are devoted to cattle. We will avoid raising cattle for meat. (Our meat production will come from the small animals that spend part of their time in the edible forest, and that require fewer energy resources.) The cattle operation on the farm, then, will be dedicated to raising cattle for milk production, even if it incidentally produces meat (calves, cull cows). Since we have eliminated grain from the cattle diet, we will return to a traditional form of cattle farming, enhanced by advances in modern agriculture:

- A wooded meadow orchard system is created.
- The mixture of perennial grasses in the grassland is carefully designed to allow the cows to be fed with grass all year.
- Hedges that crisscross the grove are composed of wood species that can provide forage to supplement the grass (animals particularly appreciate ash) as well as shelter from inclement weather. Hedgerows provide many other services, too.
- Because they do not feed in exactly the same way, it is possible to raise the same number of sheep/goats and cows together (each needs 1 hectare/2.5 acres on average), making best use of resources. Our dairy farm therefore amounts to twenty cows and as many sheep and goats.

- A small number of pigs or poultry can be intermingled, as these animals benefit from whey, a protein-rich by-product of dairy production.
- Tall stem-fruit trees in the pastures produce, in the case of apples, 10 to 15 tons of fruit per hectare on average (or 200 to 300 tons of apples annually for the entire farm).
- Hedges may be partly planted with fruit trees, like apple, plum, cherry, and pear.

AQUACULTURE

If they do not exist already, pools and ponds are dug wherever the ground is right. These numerous water holes are used to provide water to the animals and irrigate crops, but also for aquaculture. They can also be used to sanitize the waste of its inhabitants — the waste of small animals, scattered across the territory, will in turn fertilize the croplands.

MICROFARMS

That leaves 15 hectares (37 acres) that will be devoted to as many microfarms. We have already discussed the wide range of products that can be produced on a microfarm, so we won't repeat that here. But it is helpful to note once again that, within microfarms, densification of crops frees up space for fruit trees, hedges, forests, gardens, grains, vines, and the rearing of animals on a small scale (sheep, goats, pigs, poultry, rabbits). This further reduces the share that the agricultural system must devote specifically to grain and livestock.

A Virtuous Spiral Combining Ecology, Economy, and Social Life

We have taken a general look at the various productions that could make up our agrarian solidarity system. Are you satisfied with what we've accomplished? Producing such a quantity and variety of food on what was only a medium-sized grain farm may seem surprising (or suspect, in the eyes of those advocating industrial agricultural models). Our agrarian solidarity system is truly a profound break with the current practices. Industrial agriculture works mechanically, linearly, while our agricultural system is powered by people, animals, and plants. It is holistic and interwoven. It is not at all the same thing!

What is also very innovative in this vision is the role of the edible forest and the ubiquitous fruit trees among the crops, traditional orchards, and hedgerows. This calculation is to be refined but it is estimated that 60 to 70 percent of the land is wooded. In this form of eco-culture, the first task is delegated to the trees, which provide high levels of production while ensuring increasing self-fertility in the system.

But the interest of the system goes much farther — beyond agricultural production to the social component. To support the activities mentioned above, about thirty agricultural jobs have been created (about one job for every 3 hectares/7.5 acres). These farmers will in turn generate other jobs from diverse and related activities.

CRAFTS

One or more artisans can put their skills to work building and repairing the tools of the farm, and building and maintaining the lodging and infrastructure. Other craftsmen can work with materials generated by resources in the agroecosystem: weaver, potter, wood turner, joiner, carpenter.

ANIMAL TRACTION

One or two experienced leaders and some strong draft horses with the best modern equipment may offer their services to farmers according to their needs. There will therefore be virtually no need for a tractor. This transition is facilitated by the small size of the farms.

ENERGY

A biogas plant can be fueled by biomass from the forest, hedges, and various vegetation. The resulting energy can supply homes and community facilities. A logger can also produce firewood from the many trees. The proposed firewood production will be at least 400 cubic meters (110 cords) per year (10 cubic meters/2.8 cords per hectare per year) when the edible forest reaches a certain maturity, enough to fully heat forty well-insulated homes, which corresponds to the potential number of agricultural and other jobs created in an agrarian solidarity system.

FOOD PROCESSING

The possibilities are numerous. The small processing units, and milling and drying equipment, can be powered by solar, wind, or water.

SALES

The many products integral to the agrarian solidarity system can be distributed to local populations but also sold on site in a well-stocked, attractive farm shop.

RESTAURANT

An organic restaurant can operate using the local products created by the network of farmers.

ADMINISTRATIVE

Secretarial, management, and promotion services for different products and artisans can be pooled and carried out on site by a dedicated team. Over time, a coordinator will likely be needed to organize all these activities.

AGRITOURISM

The agrarian solidarity system can respond to the need to reconnect our contemporaries with nature and accommodate many visitors. Accommodations such as tree houses and eco-lodges can house them. A seminar center would find its place in this innovative space.

EDUCATION

Such a system represents a concentration of knowledge, so naturally it would become a place where that knowledge can be further developed and shared. Community members can offer public presentations to diverse audiences, including schools. Requests for future training of farmers will arise, and the agrarian solidarity system can be an exceptional laboratory to meet this emerging need.[12]

SHARED WORK SPACE

Some of the people working on the site will practice multiple activities, making shared work space a sensible option. Shared artists' studios, for instance, could allow artists to improve their economic performance by sharing a secretary, meeting rooms, and facilities. Or imagine:

Hello, may I speak to Dr. Pepper?

He is out in the field planting onions, can I take a message?

Participatory and Decentralized Governance

This initial endless inventory suggests the vast potential of an agrarian solidarity system. Forty people could work on the site. If they live there with their families, they will form a genuine ecovillage (hence there will be a need for nurseries and other services).

Again, suitable governance will need to be established. This is not insurmountable; the models are numerous. Around the world, millions of organizations, companies, associations, and NGOs manage the work of people with varied and complementary activities effectively. New modes of governance and exchange including nonviolent communication and sociocracy can enrich and humanize older models. Governance will be collective, democratic, and decentralized. An ethical charter and internal regulations may define the founding principles and unite individuals. Each actor should enjoy a large degree of autonomy in his or her activities. Personally, I think it is important that this type of system is based on an entrepreneurial model rather than a community model. When the work relationship is well structured and efficient, when remuneration is fair and reflects the contributions of everyone, there is a solid foundation for building frienships and living well together. The reverse is rarely the case.

Making New with the Old

These lines are a bottle thrown into the sea with a wish that we hope will inspire experts, research organizations, landowners, and local authorities to seize the issue and fund a comprehensive study. Forgive the loose nature of this concept, which is still only an idea. Note, however, that we did not pull this idea out of a hat: This type of agrarian solidarity system has existed in the past, even if it was not designated as such, and still exists in various forms in many parts of the world.

Until recently, France was living without fossil fuels and half of the working population comprised peasant farmers. Agrarian solidarity systems were then the norm. I visited the La Ferme Européenne des Enfants in Seine-Maritime, a small paradise created by a doctor, Veronique Barrois, in what was once a family farm of several dozen hectares.[13] Various livestock and polyculture activities were practiced on the farm historically, and they are still practiced today. I was amazed by the ingenuity with which our ancestors valued existing resources. The area is traversed by a river. Barrois showed me around the mill where flour is made from crops in the surrounding areas. The same building also housed the village baker. Wheat milled by the miller allowed his wife and an employee to keep a large chicken coop. The birds were plucked using the river water, artfully heated by recovering heat from the baker's oven. A hundred meters below, the same river fed several pools dedicated to fish farming. This gives an overview of the wealth of connections operating in a rural community of old. Our postmodern agrarian solidarity system is based on the same principles but benefits from advances in almost all areas — not to mention the conceptual tools of permaculture. If our ancestors did it, why can't we, with the means and the knowledge we have today?

Economic Viability of the Agrarian Solidarity System

We have seen that such a system raises the possibility of transforming an average farm that generates only one job into an edible landscape that supports dozens of activities. The ecological sustainability of the agrarian system seems assured, as does its resilience, and I'm willing to bet that its carbon footprint is largely positive because tens of thousands of trees will be planted and the land will not be tilled.

How many people will such an agrarian system be able to feed? A very rough estimate suggests the possibility for thirty farmers and some additional processors on 100 hectares (247 acres) to supply all vegetables, fruits, berries, nuts, grains, flours, meat, eggs, fish, dairy products, oils, honey, herbs and medicinal tinctures, mushrooms, fruit juices, cider, and beer or wine by region, with a population of about a thousand people. Obviously, agricultural systems will be adapted to the particularities of each region and will have very different outcomes.

If a thousand people are fed almost entirely on 100 hectares, this implies that 1,000 square meters can feed a person entirely. Is it realistic? It depends on the context, but seems relatively feasible in a permaculture perspective with densification of production, if the dietary changes we've mentioned have been made. A city of ten thousand inhabitants would thus need only a dozen agrarian solidarity systems to sustainably ensure its food independence.

The economic viability of the agrarian solidarity system will be facilitated by the fact that virtually all production will be sold without an intermediary, which will increase producers' income. Widespread implementation of such a system may make direct aid to producers unnecessary (under current agricultural policy, that aid represents 40 percent of Europe's budget); public subsidies could be redirected to the investments necessary to create agrarian solidarity systems and finance in-depth educational programs aimed at training new farmers.

The rise of this type of agricultural model will have economic benefits beyond helping to solve the unemployment problem. The tax burden on individuals and businesses would drop accordingly, which would boost other sectors of the economy. A new green economy would emerge, allowing us, without too much pain, to make the transition from our current industrial model to one in which agriculture and artisanal crafts form the foundation.

To facilitate this transition, our society should relocate a number of activities and services. It has become very difficult, for example, to slaughter livestock locally, since small slaughterhouses have closed down. The emergence of such farming systems also requires a shift in taxation and legislation. We cannot radically change the system of agriculture without affecting other sectors of society.

In conclusion, the potential proliferation of such agrarian solidarity systems opens up the possibility of achieving genuine food autonomy for each region, while renaturing the landscape.

Money Cannot Be Eaten

The contemporary global economy relies heavily on financial speculation. It goes from bubbles to crashes while governments and international institutions have great difficulties regulating it. Countries collapse in a few months like a house of cards, throwing millions of people into a precarious situation. The economy appears to be more and more vulnerable, and based less and less on real value.

A new economy based on an edible landscape constitutes a solid foundation for our civilization. The young Slow Money movement states emphatically: "In soil we trust," its motto.[14] The first need of a living being is to eat, and humans are no exception. The financial wealth from stock exchanges is not edible. Repositioning agriculture as the foundation of an economy that will be strong, real, and rooted is our hope of salvation in the coming turbulence.

Allow me one last argument for our agrarian solidarity system: It is profoundly beautiful and will offer a high quality of life!

TWENTY

Microagriculture, Society, Planet

I did all the calculations. They confirm the experts' opinion: our idea is unworkable. I have only one thing left to do: execute it.

— *Pierre-Georges Latécoère, founder of transcontinental air mail*

All major significant developments of humanity first were dreams.

— *Robert Blondin*[1]

s there a fourth stage to the rocket? Of course, if we want it. Pushing the same logic farther, we can imagine redesigning our landscape, radically transforming our way of inhabiting the earth.

The vision that drives Perrine and me is the creation of a million micro-farms in France in the coming twenty-five years and an additional two to three million by 2060. This date, as we have mentioned, is when there will likely be not a drop of oil left. We have anticipated this shortage by envisioning an eco-culture based on solar and human energy — allowing our society to, by then, attain sustainable food autonomy.

Again, prospective data shared in these pages are just a hunch based on what we see in our gardens. It is normal to assume that developing human activities and respecting the environment are contradictory undertakings. The abundance manifested in our gardens invites us to transcend this divide. What's good for the earth can be good for humans, and vice versa. This simple observation allows us to imagine a completely different future. Of course, critics will easily challenge each argument. We ask your indulgence and invite you to deploy your imagination. Allow yourself to dream.

Enrich our vision with yours. "A map of the world that does not include Utopia is not worth even glancing at," wrote Oscar Wilde. Let us walk together toward Utopia.

Legal and regulatory constraints should not stop us: It is not a goal, of course, to work outside the laws, but regulations and laws are at the service of humans and will adapt to future changes in society.

We switch back and forth from one world to another; the world of tomorrow must be visualized before it can be quantified. If we determine, together, a desirable direction, it will be easy to find design offices, multidisciplinary teams of experts, and PhD students to compile data and conduct prospective studies, with cost analyses and rationales, consistent and well documented. Our present society is not lacking in numbers but rather in visions, perspectives.

We are aware that a thousand nuances will arise. Many related topics deserve to be addressed. But in the limited space of these pages, we are not necessarily positioned to do so. We prefer to cultivate a holistic view and leave it to better-informed people to dig deeper into each sector.

Recall that we are up against the wall, called upon to quickly divide our environmental footprint by three. It is urgent to stop global warming. Positioning a new form of agriculture, eco-culture, as the foundation of the economy and society of tomorrow can enable us to achieve this. But our society is so cut off from the land that hardly anyone still imagines agriculture to be a remedy. While in many areas of the world a young farming movement is under way, working the land is still mostly perceived as an activity of the past. Something is amiss. Industry has trumped agriculture since the industrial revolution and we cannot seem to change our ways. In response to our present crisis, successive governments talk about re-industrializing France — as if energy and raw materials were inexhaustible, as if the biosphere could endure such attacks for much longer. It is a short-term view, a view stemming from twentieth-century thought. By seeking to perpetuate past solutions rather than looking to the future with new eyes, we keep ourselves in the wrong century. Period. The collective imagination is clamped; it would be wise to expand our perspective.

We are farmers talking about the place where we find ourselves. Experts and scholars, pardon us; we are neither of these. But we do interact daily with the earth. I have muddy boots as I sit at the computer. This book was written by night, once the work in the gardens was complete and the children were in bed. It is not a scholarly work. But perhaps it is this that allows us to envision a way forward that our elites do not perceive.

How Many Farmers in 2060?

Our intuition tells us that three to four million microfarms would be needed to cover all the nutritional needs (apart from exotic products) of a population of seventy million people in France. We can estimate that to achieve this goal of food self-sufficiency, between three and five million agricultural workers would be needed. Each farmworker would feed, in this context, approximately twenty people.

Here are some numerical benchmarks on the development of the agricultural sector in France.

NUMBER OF FARMS IN FRANCE:[2]

- In 1955: 2.3 million.
- In 2010: 0.49 million (-26 percent compared with 2000).

NUMBER OF FARMWORKERS (FULL-TIME EQUIVALENT):[3]

- In 1955: 6.2 million people.
- In 2010: 0.75 million people.

SHARE OF ACTIVE FARMERS IN THE POPULATION:[4]

- In 1900: 50 percent.
- In 1955: 31 percent.
- In 2010: 3.2 percent.

DESTRUCTION OF JOBS IN AGRICULTURE:

- Since the early 2000s, approximately 10,000 agricultural jobs have been lost every year in France.
- Since 1955, 5.45 million agricultural jobs have been lost in France.

NUMBER OF UNEMPLOYED IN FRANCE:

- In 2012: 28.5 million people employed, and 3.3 million, or 12 percent, unemployed in category A (France's designation for those who have not worked at all in the preceding month).[5]

If we count all jobseekers in categories A to E (including supported employment), in 2013 France had 5.5 million unemployed, representing 19 percent of the workforce.[6]

Setting a goal of creating three to four million microfarms and as many agricultural jobs (in addition to the 750,000 current farmers) is in no way unattainable. It is simply getting closer to the number of farmers that fed France sixty years ago. But with a size nuance: The permaculture microfarms of tomorrow will not function like the farms of yesterday and today. In many ways, they will come closer to farms before mechanization, but they will be much more productive due to the bio-inspired approach.

It is striking to note that, since 1955, 5.45 million agricultural jobs have been lost in France, as our country now has 5.5 million unemployed. From this perspective, 1955 was an interesting time because it marks the change-over from a traditional agricultural sector to a productivist model. We ask if there is any sense in subsidizing a largely industrialized agriculture that destroys many jobs and has a negative impact on the biosphere and human health, while supporting a large number of job seekers, who live, most often, very poorly in this precarious situation. Even if this idea is not politically correct, creating a large number of agricultural jobs seems beneficial for people who could access dignified work, for society, and for the planet.

The fact that an increasing number of individuals will produce all or part of their food should be kept in mind when considering future scenarios. As mentioned, we will witness a partial de-professionalization of agriculture.

And How Many Artisans?

The revival of manual agriculture will be accompanied by a revival of craftsmanship. The decades ahead will see the gradual decline of large farms and factories — perhaps even their disappearance. This will generate a large number of job cuts, but the parallel rise in microfarms and artisan workshops should absorb the unemployed. The energy descent will destroy some jobs — all those linked to professions with voracious energy consumption — but it will create many more, because it is the use of fossil fuels and the development of thermal and electrical machines that rendered human hands less necessary.

This raises the question of the desirability of farm trades and crafts. This point is crucial. Our imagination is limited by visions of the past. Everyone, of course, is not attracted by so-called manual labor — which constantly requires, however, our analytical skills, adaptability, and creativity.

But if those who work in industry and the service sector realized that there is great happiness to be found living largely outdoors, making objects by hand, being your own boss, they would line up at our gardens and our studios! Soon it would appear more fulfilling, more exciting to pilot a farm with creativity, to make an alliance with the forces of nature, and to live free in a lovingly gardened territory than to be stuck in a city, in crowded public transport or an air-conditioned office. We have the feeling that this change would begin to take hold. It will accelerate when the scarcity of fossil fuels raises the price of food. When food becomes scarce, when we emerge from the domination of plastic, the knowledge of the farmer and artisan will become infinitely respectable.

In the case of a food shortage, people in cities will court us — or they will murder us to loot our gardens. But they would do well to learn to garden themselves before we get there!

Redraw the Landscape of Europe and Beyond

Returning to our vision: If microfarms, the ecosystems of microfarms, and solidarity agricultural systems become widespread, we could give a completely different face to our country.

Being able to concentrate crops in substantially smaller areas is one of the finest prospects offered by permaculture, because a field of possibilities opens up. It frees up space. Again, some figures are required:

TOTAL AREA OF METROPOLITAN FRANCE:

- 552,000 square kilometers, or 55.2 million hectares (136 million acres).[7]

AGRICULTURAL LAND USED:

- 27 million hectares (67 million acres), 50 percent of the territory.[8]

FOREST AREA:

- 16.4 million hectares (41 million acres — an increase of 0.6 percent per year).[9]

The spread of bio-inspired agriculture would make immense areas of agricultural territory available. The most relevant use of these freed-up areas, in a permaculture vision, would be to plant trees by the billions. "To make up for the loss of trees in the past decade, we would need to plant 130 million hectares (or 1.3 million km^2), an area as large as Peru. Covering the equivalent of 130 million hectares would entail planting approximately 14 billion trees every year for 10 consecutive years. This would require each person to plant and care for at least two seedlings a year," according to the United Nations Environment Programme, which launched a campaign to plant a billion trees.[10]

Let's examine the consequences of such a transition.

- If four million microfarms — covering on average 3 hectares (7.5 acres) each (taking into account the areas that devolved to grain crops and livestock breeding) — can feed seventy million people, only 12 million hectares (30 million acres) would be needed to cover the food needs of France.[11]
- Since our bio-inspired model gives priority to trees, at least 60 percent of the cultivated area is wooded. So trees would cover 7.2 million hectares (18 million acres) of the 12 million.
- These 7.2 million wooded hectares would be added to the 16.4 million hectares (41 million acres) of existing forests; the forest area would then amount to 23.6 million hectares (58 million acres).
- If 12 million cultivated hectares is enough to supply our food independence, it would be possible to release about 15 million hectares (37 million acres) of agricultural land, more than a quarter of the national territory, for other uses: edible forests, trees dedicated to wood energy and timber harvest, orchards, meadows for breeding draft animals to promote horse-drawn transport over short distances, biodiversity reserves.
- Biomass from the forest can replace petrochemicals, confronting a key challenge of the bio-economy.

In such a scenario, the France of 2060 could therefore largely double its forest area, from 16.4 to 38.6 million hectares (95 million acres) of forestland — more than half of the national territory. The country would have recovered its green canopy from the Neolithic period, but the trees would now shelter a large number of human beings who have finally learned to live in peace with the biosphere.[12]

A total of 38 million hectares of forest for a population of seventy million people suggests that each person would have about 0.55 hectare (1.36 acres) of woodland (to which must be added the private and recreational gardens and grounds and woodlands included in urban areas, which are not included in this estimate). This surface would be sufficient to absorb our greenhouse gas emissions — given the fact that our emissions would have decreased significantly with the evolution of our lifestyles. Cropland would also become carbon sinks, whereas now the vast majority of agriculture is heavily emitting greenhouse gases. So we would probably, finally, be able to collectively achieve a positive carbon balance, which seems unthinkable under the various current projections, including the more audacious ones. It should be noted that once reconstituted soils and forests return to balance, carbon sequestration will cease. Repairing what coal, gas, and oil have emitted as CO_2 will require new levers.

Entering the Era of Solutions

Such a transition would be capable of generating a powerful virtuous spiral whose beneficial effects could help solve many contemporary problems. Let's summarize some of the potential.

GLOBAL WARMING
Cropland and forests should be able to absorb a significant portion or even all of our greenhouse gas emissions.

EMPLOYMENT
By creating three to four million agricultural jobs, and including incidental jobs and many artisanal jobs, the growth of microagriculture could help eradicate unemployment. Unemployment is the scourge of a declining industrial society, and one that seems impossible to contain. All successive governments make unemployment their priority, with results that we know all too well. Maybe it would be helpful to take a step back and consider the

recent history of France. In the early twentieth century, a policy choice was made: favor industry over small-farm agriculture. By undervaluing agricultural products, we freed purchasing power that was then applied to industrial products. By deflating the agricultural workforce, workers were available for the factories. A rural exodus and swelling cities led to a centralized policy and a certain vision of "progress."

We talked about the 5.45 million agricultural jobs that have disappeared in France since 1955. These lost jobs correspond to the number of currently unemployed. Yet no one has made the link between the jobs destroyed and the jobs now missing.

By creating a large number of agricultural jobs, we could probably return to a form of equilibrium. The cost of creating a microfarm (from 50,000 to 100,000 euros/$56,500–113,000) is equivalent to the average cost of two to three years of government support for an unemployed person. Yet a microfarm is a working tool that remains useful for many generations. It would make sense to invest heavily in the creation of many microfarms, which would enable those who suffer from unemployment, and who would be drawn to a life close to the earth, to find a beautiful career and a new focus for their lives.

Training programs could help workers meet meet expectations. Working the land is now completely absent from our educational programs, except for a few specialized courses.

HEALTH

Widespread microfarms would offer everyone an abundance of organic food. This societal change would greatly contribute to addressing the famous deficit of the social welfare system because organic food, alive and local, would improve the level of health for all of our citizens — not to mention the fact that growing food makes an excellent outdoor activity for all those who want to reconnect with the earth.

ENERGY

Reforestation would allow augmented use of wood energy, in addition to other renewable energies. This would facilite the phase-out of nuclear power. However, it would be wise to improve combustion technologies to diminish particulate emissions.

ECO-CONSTRUCTION

Forests produce great quantities of materials for green building. An increase in the construction of wooden houses, whose health benefits have been demonstrated, would help to limit the frightening amount of concrete, a heavily energy-dependent material used today.

ORGANIC CHEMISTRY

The development of biomass could replace petrochemicals.

BIODIVERSITY

Devoting millions of hectares to biodiversity reserves would balance the development of human activities and the preservation of wildlife. Recall that, in permaculture, wildlife and wild flora can thrive in cultivated areas.

ART OF LIVING

We have no pictures or videos of what France will look like in the year 2060, but wouldn't it be great to live in a country that is lovingly gardened and partly restored to nature and rewilded!

WORLD HUNGER

Subsidized exports of agricultural products to the Southern Hemisphere are destroying local agriculture there. Similarly, farmers there who grow food for export to northern countries are often deprived of their own food crops. Setting a target for each country to achieve sustainable food autonomy and north–south cooperation would do much to limit hunger in the world.

THE INVENTION OF A NEW WORLD

Agriculture is closely linked to other spheres of society. Choosing to base the civilization of tomorrow on a bio-inspired agriculture will profoundly alter our lifestyles.

We have seen that rural landscapes will be completely redesigned. It would be the same for the distribution of our habitats. The emergence of a large number of microfarms and artisan workshops would create migration from cities to the countryside. Peasants and artisans would form new hamlets or swell existing villages. The latter are mostly in decline.

The Future Is Local

"Men are like apples: The more we pile up, the more they rot," asserted Mirabeau.[13] The ever-growing concentration of population in megacities will not survive the energy descent. Should we regret it? "Our species is not made to live in large herds, but in small cells. All our social theories should take this principle into account," noted the zoologist Desmond Morris.[14]

The return to local living that we are calling for will be accompanied, when it is established, by a return of schools, small shops, and other services like public transport, post offices, and health care to small urban areas as new residents and new families arrive. Cultural life would find a new dynamism. The increase in the population of small towns and villages will increase farmers' customer base and grow the community of artisans.

Since a growing number of goods and services will become available locally, and it will become easier to find a job close by, the movement of goods and people will be reduced considerably. Currently the average daily travel of a French citizen is 25 kilometers (16 miles) for a duration of sixty-six minutes.[15] What a waste of time and energy! In the world of tomorrow, most journeys will be made on foot, by bicycle, or by local public transport (which can be, for some, horse-drawn). This change will reduce the significant carbon cost of transporting people and goods, which represents 19 percent of our footprint and remains one of the most difficult emission reduction problems to solve.[16]

The redistribution of the population driven by the expansion of small farms will result in a new norm for urban areas: The standard will not be the big city, but small and medium towns and villages, interconnected to larger urban centers by new short-distance communication networks.

This transition will limit very substantially our environmental footprint: Our food, our transportation, our homes, our consumer goods will see their energy and materials costs severely reduced. It will become possible to attain the only viable option: using only one planet to cover all our needs.

The Global Village

I see some readers raising their eyebrows: Return to the village, the small town? What imprisonment! The provincial life of yesteryear was often warm and supportive, but also limiting, exposing us to the scrutiny of others, to a certain conformity from which large cities set us free.

But the village communities of tomorrow will be very different from those of the past. The energy descent that begins in the coming decades will severely limit the movement of people and material goods. We no longer will jump on a plane to spend the weekend in New York and eat beans from Kenya. In this, our lives will be more like those of our grandparents. But today and even more so tomorrow, the circulation of the intangible goods of humanity will be global. New modes of communication have revolutionized our relationship to the world. Technical, cultural, artistic, and spiritual changes are becoming universal. Every village in Europe can share with other villages from all continents. On condition, however, of finding green technology to manufacture computers and run the Internet (or the future iteration of these communication tools).

The village community of tomorrow will be connected to almost all other communities around the world. It will be possible to live rooted to the land and have an extended family, all while deploying our potential to be human.

We can also hope that we will substantially reduce the level of violence in the human race. The emotional and material security provided by the village community (the famous visceral needs of human beings to live in a clan finally sated!), and the unprecedented opportunity to spread our wings thanks to the expansion of local activities and global sharing of the cultural and spiritual treasures of humanity, should allow us to develop more fully as humans. Violence and envy will have less reason to be exercised once basic needs are covered, social disparities are reduced, and everyone can thrive and share in a supportive environment.

Toward a True Democracy

Expansion of local communities will be reminiscent of the golden age of ancient Greece, the cradle of democracy. The vision shared in these pages can only be embodied with a push toward a decentralized form of democracy. Currently, in France, we live in a relatively centralized form of democracy — which is a step in the right direction. But the ecological transition, the permaculture approach does not lend itself to centralization, which

better suits the industrial model. We must achieve optimal levels of autonomy and freedom in order for every citizen to begin implementing new
solutions in terms of food, energy, eco-housing, health — the solar panel
rather than nuclear power, the family garden instead of the supermarket,
lumber from the nearby forest rather than Home Depot, medicinal plants
instead of big pharmaceutical plants. In other words, the transition should
release the creative genius of civil society, as proposed by Pierre Rabhi.[17]
This, incidentally, is a powerful invitation to revisit our educational system.

Tomorrow, each local community will take its destiny in hand. The
principle of subsidiarity, which involves giving a maximum amount of
decision-making power to the lowest possible level of authority, will release
a cascade of creative energy. Everyone will be able to overcome the over-
powering canons of modernity, the stereotypes of happiness and
consumerist injunctions it conveys, to invent their own destiny. But all will
develop, we strongly hope, common values, starting with the urgent need
to protect all forms of life that share this planet with us and without which
we cannot live. The post-oil world will be at once more rooted in local life
and more open to the universal world.

TWENTY-ONE

The Earth
Is an Adventure

Our deepest fear is not that we are inadequate, our deepest fear is that
we are powerful beyond measure. It is our light, not our darkness,
that most frightens us . . . Your playing small doesn't serve the world.

— *Marianne Williamson*[1]

We will be leaving you soon. Let's go back for a few moments to La Ferme du Bec Hellouin, if you like, before thinking about starting a microfarm in this day and age.

We have related, through these chapters, how we were inspired by forms of farming deemed to be the most productive and sustainable in order to arrive at our own synthesis, with permaculture as a conceptual framework. Microagriculture — or permaculture, or eco-culture — is still in its infancy, yet it is already proving rich in so many ways that it opens an important path for the future of the world. It allows generous harvests while improving the environment, and requires only a few tools and little or no oil for its implementation. It raises the possibility of creating many jobs and beautifying our environment while helping to eliminate some of the scourges afflicting our societies. This form of eco-culture suggests that it is possible, by going with the flow of life, to feed the people while healing the planet.

The Experimental Ferme du Bec Hellouin

Our farm has become a sort of open-air laboratory, a beehive where many projects move forward, often in haphazard ways, held together with bits of string — yet everything progresses from one year to another. Maintaining cohesion is often a balancing act because the tasks are varied and the demands of our work are completely disproportionate to the time and resources we have. The farm receives hundreds of requests for projects and training from municipalities, associations, media, individuals, and companies. François Léger is inundated with applications from students, colleagues, and local authorities. This interest in permaculture is gratifying and helps us move forward because it leads us to experts with possible solutions, with wonderful knowledge to share. But it also pushes us to our limits and takes us away from the everyday work that is the foundation of our business. Our daily struggle is to free up our time to work in the gardens! Staying focused, avoiding fragmentation, and avoiding becoming overwhelmed is a challenge in asceticism. For Perrine and me and the rest of the team, the pressure is such that it could make us explode mid-flight.

The full-time team comprises twelve people, spread among agricultural production, research, and training. Each of these three divisions is managed by a dedicated structure, and only the research, led by the nonprofit Institute Sylva, is subsidized. The other activities are self-financing. Vegetable production, although it is experimental, still has an expected economic contribution, which requires us to be constantly in tune with the realities of the profession. Daily, the full-time staff form one team that works side by side and spends time on each division. Trainees working on master's or doctoral degrees complete their studies here.

Our common desire to embody the ethical principles of permaculture in our work is a real one. The pay is almost all equivalent, regardless of the function: The team includes an engineer and some who have no college training at all, but everyone is very knowledgeable and invested, and shares the love of work well done and camaraderie.

Overwork and dispersion are our main obstacles. Stress, the poison of modern times, is a daily visitor, even deep in the gardens. A close friend, Sébastien Henry, a business leadership trainer, gave us a beautiful gift this year: He offered to coach our team to help us survive the adventure as passionate and demanding as when we started.[2] Sébastien had us reflect on the mission of the farm, and the following mission statement emerged from our exchanges: "La Ferme du Bec Hellouin's mission is to explore innovative agricultural practices to feed the people while healing the Earth. To do this, it is open to exchanges while remaining a cocoon for those who work there." This cocoon reference deserves an explanation. Sessions with Sébastien high-lighted the fact that we have a lot of fun working together — but excess stress is affecting us all. There is a balance between openness to the world and internal cohesion. We have consciously made the choice to make gardening our priority, since it is based on this that the other two poles of the farm, research and training, rely. This leads us to sometimes close the doors of the farm, which some find hard to understand, but it's a matter of survival.

Staying Market Farmers: Our Challenge

Remaining market farmers is very important to us. It would be easier and more lucrative to develop training or work as consultants, but this is not our vocation; our passion for farming is growing year after year as we delight in each day spent in the gardens. We have much more to learn and are very far from the level of excellence to which we aspire. Eliot Coleman has said that working in agriculture is like climbing a mountain, only better: It has no top. The list of study topics continues to grow and we are currently working on the following projects: the design of a bioclimatic greenhouse and innovative tools, perennial plants, crop mixings, auto-fertility, forest gardening, integra-tion of animals in the crop system, microorganism cultures, and more.

The discomfort of our situation is the fact that there are currently too few players with experience in the field of permaculture microagriculture. We aspire to create an international network of people and place resources. But how to train these human resources? The issue cannot be debated in depth in these pages, but it is at the heart of our concerns. We realized that we are too small to generate a significant number of trainees here and have chosen to prioritize the development of training materials that can be used by a greater number of people.

The book you have in your hands is the first of these supports, dealing with concepts of natural farming. Our market gardening manual is in the works; it will complement this first book by describing, in very concrete ways, how to apply these concepts in a garden or on a farm, offering a lot of new information and tools allowing individuals and professionals to create bio-inspired agroecosystems. We have also created a series of educational films that is accessible to all through our website www.fermedubec.com.

Recall that the collection of the Bec Hellouin school of permaculture now offers hundreds of freely available documents.

Our dream is to exchange information and techniques with farmers from the Southern Hemisphere. We collaborate with several NGOs and sometimes host, at the farm, teachers that work with farmers in Africa. We try to trust life. Instead of dreaming about a slowing down, which we probably will never do, we try to learn how to create calm in our lives, even in the midst of the agitation. Again, we are far, very far from our goal! We have hardly a spare moment — yet last week, for the first time in six years, we took two days for ourselves. It was great, but we decided that two days were enough. More vacation time would bore us. Perrine and I agree that we will work until our last breath if life allows us to. But can we still call such exciting activity work?

Creating a Microfarm Today

It is surprising that in just a few years the adventure of La Ferme du Bec Hellouin became so intensely followed in the world of agriculture and beyond, prompting strong mixed reactions. Many people do not believe the results, but generally they are poorly informed of what is happening here and have not read the interim reports of the study. An increasing number of people also feel inspired by the concept and wish to start their own micro-farm adventure. Currently in Europe, to meet the growing consumer demand, the trend is to move toward industrialization of organic farming. Many agricultural stakeholders will not find a place for themselves in this trend. The echoes that come to us from all corners of France show that permaculture brings a hope for an in-depth renewal of organic farming practices, in the direction of more natural farming and more suited to the world of tomorrow.

More than a Job, a Lifestyle Choice

What can we say to those who want to create a microfarm? First, being a market farmer is more than a job, it is a lifestyle choice that imposes heavy constraints, while offering undeniable opportunities. The constraints are real, even if they vary depending on the business model. In a permaculture approach, we seek to create a complex agroecosystem; implementation and management require a strong personal investment and the acquisition of a vast bank of knowledge and skills. It will probably be necessary to say good-bye to the holidays, for a few years at least, and agree to live with very little money — even earn nothing at all at first. A lifestyle choice as radical as this has a strong impact on all aspects of our lives, but also on our surroundings and our families. It is necessary to have the support of close relatives. Do not embark lightly on such an experience!

But the rewards are on par with the personal investment and the risks — provided, however, that obstacles are overcome and the farm becomes sustainable! For us, living in nature is real life, a varied and ful filling existence that develops our potential as human beings. The reward is not money, nor is it social recognition, for agricultural work remains unjustly devalued. The reward is in the quality of life. The wonder at the daily spectacle of nature, the morning mist, dewdrops, the magic of seed germination, bird companions, the beauty of vegetables and fruits, the joy of eating our own products and having real and sincere exchanges with honest people: All this is priceless. Working from home, being your own boss, living outdoors, making choices and accepting the conse-quences makes for a noble and full life, although the concerns are real. The days are long but, as John Seymour put it, when the evening comes we ask, "Done already?"

Becoming a farmer also offers the satisfaction of committing our lives to the service of the earth and beyond: Every movement has meaning and can contribute to the common good. Finding purpose in life, in a world charac-terized by the loss of meaning, is a privilege.

And for the family, even if we do not go on vacation, even though week-ends are partly devoted to the farm, what a gift it is to live together in the heart of a large garden! Each of our girls has her animals (sheep, rabbit, chicken, pony, dog, or cat), a home base with an infinite variety of games, trees to make a treehouse. Our children are growing up in a small intensely alive and safe world where they can explore their capabilities. Yesterday Fénoua, our fourth daughter, joined me on the island after school and said, "Dad, I smiled at the ducks and one of them smiled back. He is so nice, this one is my duck!" Growing up on a farm with a smiling duck — that is the childhood we wanted to offer our children. The small miracles mentioned at the beginning of this book are exponential!

Often visitors who see the work accomplished say, "What courage you have!" We don't dare to tell them that, for us, the real courage would be to live in the city, taking the subway or enduring traffic jams on the highway to go shut ourselves up in an office surrounded by synthetic materials, with air-conditioning and the noise of the coffee machine, awaiting our paid leave. Each life has its constraints. I prefer a thousand times that of the farmer. Lack of vacation is not that serious because we are at home.

The Economic Success of a Microfarm

But to be able to enjoy life on the farm, you must not be overwhelmed by worries. For the beautiful dream not to veer into a sustained nightmare, it is important to carefully prepare every aspect of the farm's design and construction. Unanticipated things will arise no matter what. When organizing expeditions aboard the *Fleur de Lampaul*, I often repeated to the crew: "The crazier the dream is, the more seriously we must prepare." And I added, "We will try to go as far as possible, but we're not going anywhere without a plan." This state of mind allowed us to navigate twenty-two years without an accident, and it seems well suited to farmers, too, who are also adventurers in their own way.

So prepare your project meticulously. Give yourself time — several years. Move ahead gradually. Start by cultivating a garden, putting yourself in the skin of the professional you aspire to become, but on a reduced scale. Give or sell your production to your neighbors. Keep gainful employment as long as possible, until you have acquired sufficient expertise to take the plunge.

When will you be ready? We strongly advise you to quantify the production of your garden, its economic value, even if only virtual, the time it will take to produce it, the costs to be incurred. For a market garden, when you reach 25 euros ($28) of vegetables per square meter per year, on average, and when each hour of work will generate at least 15 euros ($17) in sales (for all tasks, including maintenance, marketing, management), you can estimate that you have some chance of earning a living (modestly) in the vegetable business. Both figures seem to represent the threshold beyond which a manual microgardening operation becomes economically viable — adjustable for every person and every context, of course. You need a good level of experience to achieve this. The combination of these two indicators reflects, among other things, the technical skills of the market gardener, but also the speed at which work is completed — very important in this profession.

Train yourself as much as possible before you start, rather than after. Our friend Jean-Martin Fortier believes it takes ten thousand hours of work to become a market farmer, and this number jives with our experience: This is the fifth year that we have begun to feel at ease confronting the complexity and diversity of tasks. Time invested in training is time gained and money saved. Remember: Permaculture farming is agricultural knowledge. You will not need sophisticated and expensive tools; it is in yourself you should invest!

Given the virtual absence, today, of training for someone who wants to start a market-scale permaculture farm, we advise those who ask us about training to take classes in market gardening (earning an agriculture degree) and to complete the course load by personal reading and research, internships with different professional farmers, and diligent practice in a test garden. Following this with a certified permaculture course seems essential.

We are often asked, as mentioned before, about the issue of initial investment. It is almost impossible to give a reliable answer today, given the current lack of references and the fact that installation costs vary depending on the business and contexts. We have estimated the average start-up cost for the creation of a vegetable microfarm to be in the range of 75,000 euros ($84,750). You can start with much less: A small garden and raised beds can be created at a very low cost. The budget of the installation, however, should also include enough to live on for the first few years, until the business takes off. All the time invested in your training will be money saved when creating your farm. Avoid unnecessary investments and your farm will also become productive and profitable faster.

Surviving the Early Years

The first few years are the toughest to get beyond because the new market farmer is confronted with many difficulties: a lack of great technical mastery, soil that is not yet fertile, heavy weed pressure, the necessity of building a sales network. Be aware of personal investment and energy needed to get through this turbulence! Your feathers will be ruffled and you will need to be supported.

It would be fair, as we've mentioned, if local communities and society as a whole supported new market farmers in the initial period during which they are so vulnerable. The services they will render when stabilized more than justify the cost. It is clear that this level of public support is still a long way off, however. State aid for organic farming is totally inadequate. Despite the facade of ads put out by successive governments, organic farming remains the poor cousin of French agriculture. At the level of local communities, as awareness begins to emerge, we suggest to new farmers that they fully integrate with the local social and economic fabric. Many municipalities aspire to have organic farms in their region. Working hand in hand with local communities is a key to success.

We strongly emphasize the difficulties because reality is not rosy. An estimate circulating in the profession suggests that the French organic market gardener earns an average of 800 euros ($904) net per month for fifty hours of weekly work and an investment that often exceeds 100,000 euros ($113,000). What other profession is as badly off? It could be said that the organic market farmer is exploited by society. We hope that the importance of this art will be better perceived in the coming years. Our friends at the Sainte Marthe farm talk about the ethic of market gardening. But meanwhile, how to get out of the relative insecurity in which most market farmers find themselves? If they have no control of the macroeconomic environment, farmers struggle to make progress in their practice to become more efficient. This is what we constantly seek at La Ferme du Bec Hellouin — not to announce beautiful figures in the reports of our study, but to help our colleagues gain in security and quality of life. The study conducted at our farm demonstrated that it is possible to earn 1,500 euros ($1,695) net, or more, monthly, by working at home in the garden, the trade of organic gardener will become more attractive.

Let us not forget that a permaculture farm produces more than food. A day will come when other services rendered to the community will be recognized and valued.

Convert Your Garden into a Microfarm

The best way to begin a market microfarm is undeniably at your home, in your garden — if you're lucky enough to own one. We get many people in our training programs who struggle to acquire a few hectares, while they live in a house with a plot of 1,000 square meters or more. Look no farther — your microfarm is there before your eyes! And if, in addition, your home is located in the city or in the suburbs, you have an incredible advantage: proximity to your future customers. The positioning of the microfarm significantly affects its success. It is much harder to sell products when you're located in open country: You sell your products cheaper, at more effort. By creating a microfarm home, you will not have to move, your spouse can retain his or her job, your children will stay in their school with their friends, and the transition will be that much easier. Transforming a few hundred square meters of lawn into a hyperproductive garden with virtually no monetary investment can be very pleasant, efficient, and helpful, as you are already inserted into the local fabric.

Do not yield to delusions of grandeur. Keep in mind the importance of limiting the investment and operating costs to a reasonable level. Avoid debt as much as possible. Better to achieve less revenue with little investment, if the configuration of your system allows you to leverage what you already have. Net margin will probably be more important and you will be calmer.

Do not make the mistake of trying to grow on too large a surface: This is the main source of potential failure. Professional agricultural advisers might push you to go big, saying that you cannot make a living on a small area. But remember the council of elder market gardeners' advice: Choose the smallest plot of land possible, and cultivate it exceptionally well. If we were to start from scratch today, we would cultivate an average of 600 square meters each, no more. But with the highest level of productivity.

In the market gardening manual to come, we will incorporate the figures of the study we are conducting and provide more specific references, crop by crop. Perrine and I feel like we have been through the mill, in leaving the beaten track behind. May the experience so laboriously gained be used to facilitate the creation of a large number of microfarms!

TWENTY-TWO

Bio-Abundance

Revolt and only revolt is the creator of light, and this light can only be known by way of three paths: poetry, freedom and love.

— *André Breton*[1]

Be realistic: Ask for the impossible.

— *slogan from the general French strike of May 1968*

*W*e are organic farmers and proud of it. We find our joy in an intimate and daily companionship with plants and animals.

Like mothers, farmers bear life. The body of our customers is formed from the body of the fruits and vegetables that we grow with tenderness and respect.

Our deepest desire is to understand, from within, the dynamics of life, at work in the soil and roots, stems and leaves, legs and wings, rain and wind. By what miracle are minerals from the bedrock assembled to form, mixed with sun and water, the soft, warm body of our donkey, the blue or brown eyes of our children?

"The grain of life and the grain of thought, successors to the material," wrote Teilhard de Chardin.[2] Our job is simply to serve this movement of life, with infinite gratitude.

In the School of the Sea and of the Land

For many years, I have been trying to understand the connection between what I discovered about the dynamics of life in the heart of tropical forests and coral reefs visited in my youth, and what takes place, what is exchanged, in that which is growing in our gardens. Can we expect to approach, in our Normandy valley, something like the flamboyance found in the jungle, the exuberance of the coral reefs?

I also think about how we as humans interact with nature to draw our sustenance. Is it possible to invent a post-oil society that combines the harmony of the first peoples with the cognitive and technical advances of the modern era?

I'm not rushing to any conclusions. It took me thirty years to understand that a simple and powerful principle is at work in nature, a dynamic that inspired the first peoples. We could name this dynamic bio-abundance.

Bio-abundance is the art of doing a lot with very little. Only what's alive can achieve this.

Assembling Widespread Molecules

Biological processes have the fascinating power to assemble and organize scattered molecules, to create cells and living organisms. They use mineral resources from the mother rock, sunlight, and gases from the atmosphere to form beings that move, eat, grow, reproduce, and die in their turn to participate in the fertility cycle.

I was awe-struck when I learned that coral reefs and the Amazon forest, the two richest ecosystems on the planet, are nutrient-poor environments, yet they thrive with high levels of solar energy. These highly complex systems generate the conditions for their own benefit by pulling the maximum value from all available resources, multiplying the exchanges in symbiotic relationships. All waste is obviously recycled. These ecosystems are major contributors to the overall functioning of the biosphere.

From soil washed away by excessive rains, waters deprived of plankton, biological processes were able to create an explosion of systems with surprisingly varied forms and colors. An overabundance — a bio-abundance.

Relatively speaking, something of this bio-abundance already manifests itself in our gardens. In seeking to align our practices more closely with what we perceive as biological processes, we observe a generous and good production that does not exhaust the environment, but rather participates in its improvement. So we can make abundant crops in gardens that are becoming increasingly fertile.

Little Fish Become Big

Biological processes — and they alone — have the power to increase resources: The little fish becomes a big fish, the seedling gives way to a mighty tree, fruit produces a thousand seeds that become almost as many new plants.

The first peoples intuitively understood that they could play a role in these living resources that accrue to, without weakening, the surrounding environment — especially if they contribute to its aggradation with good practices — replanting pits, for example. They have limited their harvest to an acceptable threshold, imposing on themselves forms of self-regulation similar to those governing plants and animals.

As long as the balance is respected, plants, animals, and humans can take the resources they need, while the surrounding environment continues to evolve, slowly but surely, toward more complexity and fertility.

In other words, bio-abundance gives plant, animal, and human communities a chance to sustainably meet all their needs, as long as the harvest is in balance with the biological processes that are capable of regenerating.

The resources produced by biological processes are organic: plants and animals. Since the appearance of the first humans, these plants and animals have supported our existence, giving us enough to feed us, clothe us, warm us, move us, heal us, build our habitats, our musical instruments, our ritual objects. Over the ages and until the entry into modernity, organic resources have made almost all of the necessary resources for human communities. Only a few minerals complemented them, mostly superfluous ones — stones and iron.

But biological processes also provide us with other services — immense services — because they can strongly modify the physical environment: They exert influence on climate (rainfall, temperature) and geology (soil formation, limestone), purify water, enhance the oxygen content of the atmosphere, and sequester carbon.[3]

Taking the resources that we need by simply tapping into the natural processes expanding in the biosphere is akin to drawing upon the interest of capital without reducing the capital itself. It is wise. It is also possible to take abundant mineral resources — stones and base metals, which form the earth's crust — without significantly eroding the integrity of the planet.

We humans, like other living organisms, can take, but also give. Enjoy natural resources but also encourage a dynamic enrichment of the surrounding environment. Take part in the cycle of fertility. In other words, we can cooperate with other life-forms to increase capital, and therefore the share of interest we can use. In this respect, we should favor cooperative relationships rather than predation, commit to limiting our consumption, creating closed-loop cycles to convert all of our waste into new resources. This is what is taught by permaculture and biomimicry.

In view of the above, we could well describe bio-abundance this way: the creation of recyclable goods based on a bio-inspired use of renewable natural resources. These goods are available in quantity and sustainably. They do not produce waste, do not damage the health of ecosystems, and may eventually contribute to their amelioration.

Bio-abundance is a concept that is new without actually being new, of course. It seemed to us an important thing to describe because most of us underestimate the potential productivity of a healthy ecosystem and the possibility for humans to contribute to the growth of biological resources with careful and thorough observation of the dynamics at play, and interacting with restraint and respect within these dynamics.

The first peoples live in what ethnologists describe as societies of abundance — we could say bio-abundance. We mentioned the Wayana people: They do not store food (except in very rare circumstances), have no fear of going hungry because they know that their needs will still be covered by the surrounding nature. Respect for life is at the heart of their culture and spirituality. Their detailed knowledge of their natural environment allows them to take full advantage of bio-abundance. Modern society has turned its back on a nature that it artificializes, in turn harming biological processes and undermining the ability of ecosystems to generate bio-abundance. This transgression was made possible by the use of fossil fuels, which have helped to generate another form of plenty, very different in its essence. The current glut of consumer goods comes from an extractive economy that preys on nonrenewable mineral resources, consuming organic resources without any awareness of acceptable levels, without returning anything to the cycle of fertility. This form of abundance could be called technoabundance. It digs holes in the earth and quickly exhausts its capital. "The biosphere is a set of geochemical cycles that connects stocks of material with energy, between which exchange flows. The flow allows stocks to recover. What does the economy do? It draws from natural resources, that which has no economic value in its natural state, without worrying about its reproduction," writes René Passet.[4]

We propose the following definition of technoabundance: creating goods with little or no recyclable value based on a predatory utilization of natural resources, renewable and nonrenewable. These goods are available in quantity, but for a limited time only, and not for all. They produce waste and contribute to the destruction of the biosphere.

Living on Earth as a Poet or Murderer?

The industrial agriculture generated by our consumer society is a profound perversion of the farm profession: Rather than a supporter of life, it becomes a sower of death. Farmers are the first victims, and awareness is emerging within the profession. Industrial agriculture is only a reflection of our biocidal society: It kills living beings in quantities that exceed comprehension, it stinks of death. It is important to stop it as soon as possible, before it definitively degrades the only known living planet. Yet like a compost heap, our society carries within it the seeds of tomorrow's world. Let us remember those countercurrents that Pupoli artfully mastered. Have confidence: The earth has the power to transform what dies into new forms of life.

The postmodern society that we will create together will operate in the mode of bio-abundance. A new form of abundance.

The news is good. Most of our contemporaries do not suspect the potential of biological processes. To abandon the technoabundance model is, they believe, to sink into penury. But no! This can lead to bio-abundance.

How do we get there? We explored agricultural themes in this book: developing the greatest possible understanding of biological processes, and analyzing our interactions with the biosphere using the logic of life. It's a 180-degree change from what we have done over the past centuries in our so-called developed societies. But we have learned a lot from our mistakes, and learned much about the workings of nature. We are better equipped than ever to enter an era of respect and fulfillment of life. We just have to develop our conscience.

A Green Society

The green society described in the previous chapter is a society founded on bio-abundance. By surrounding every house, every village, every city in an intensely alive natural cocoon, giving back to the earth the covering of trees that humans have greedily stripped from it, taking great care to limit our harvest to only the resources created each year by the biological process, we can — like Japan during the Edo civilization — provide for our needs with biological rather than mineral resources. We can live sustainably on earth. We will give up our gadgets and waste, but we will be rich with the essential goods — tangible and intangible — that make for a really good life.

This is what permaculture offers, which is both a science and an art of living. We are only at the very beginning of the adventure; the rest of the story remains to be written, and each of us can contribute.

Once again, the main obstacle is not technical, it is inside us — just as the solutions are. We all have a tendency to subscribe to limiting beliefs: *The past is past, progress is always good, the West is better than the rest of the world, working the land is for the serfs, an intellectual job is better than a manual job, my car and my iPhone make me happy.* Step back from the mental formatting that we all experience to consider the question of our common future with greater perspective. Dare to imagine the new. Take the best of the many traditions of humanity, and the best of modernity, to shape a world that has never existed. Become explorers of the future.

Finally, Perrine and I wish to share a conviction that the human being has an essential role to play, a positive and constructive role in the future of the biosphere. If nature has given us a brain as sophisticated as this, it is not to destroy in return but to enter an active process of coevolution with it. We can cooperate with biological processes to create new life-forms and new forms of organization of life. Gardeners do nothing more when they change wild plants to yield flowers, fruits, vegetables. Each scented rose, each ruby-red apple, each sweet carrot is the result of a symbiosis between nature and generations of gardeners. Nature would not have created without us there flowers and fruits. But must we still position ourselves outside of nature? We are nature, her very best effort perhaps, and our mission is to ensure gentleness and wisdom with all our fellow travelers. They expect only one thing from us: that we might become truly human, that we show ourselves worthy of this unique position of ours.

Every garden, every farm can become a place of healing in the world and contribute to its embellishment.

— LE FERME DU BEC HELLOUIN, MAY 16, 2014

══ AFTERWORD ══

*F*or three years now I have worked regularly with La Ferme du Bec Hellouin. The study we conducted had a pretty amazing response. As the scientific arm of this work, I am often asked about the reality of the published results and the reproducibility of the experiment that Charles and Perrine conducted. Answering the first question is not difficult for me. The data were produced in compliance with methodological rules imposed by any serious scientific work. Their treatment is not finished yet. The continuation of work begun in 2011 will go farther and illuminate still unclear points. I will return to the second question, which is relevant and necessary. I would like to mention another issue that is related and very often addressed to me: Why are you interested in La Ferme du Bec Hellouin?

Although I am now used to it, this question always surprises me. I often amuse myself with a lighthearted answer: I'm interested in Bec because it's close to where I spend most of my weekends. It does not fool anyone, but it allows me to change the subject and get back to the more technical and easy points. I fear I cannot skip over it lightly here.

Why would a researcher be interested in La Ferme du Bec Hellouin, a unique farm whose statistical representativeness is inevitably questionable? To be honest, I have been interested in "special cases" for over twenty years. Very simply, it is from these particular farms that the technical, economic, and social innovations that will shape the agriculture of tomorrow emerge. I'm interested in Bec because Charles and Perrine and all who work there are undoubtedly innovative. Because this search for innovation is one of the fundamental reasons for their projects and their work. And also — not least important — because they consider that the success of their project is driven by a collaboration with science and the critique that science embodies.

Why would a researcher belonging to a renowned scientific institution, who has worked mainly on large-scale modernization of industrial agriculture, be interested in a farm and people whose vision is orthogonal to that

Miraculous Abundance

historic institutional project? Belonging to a scientific institution does not mean that the person is this institution. And, anyway, this institution has long accepted that critique of any aspect of this "historical program" opens wide the opportunity to explore other perspectives for those who wish to do so in a scientific manner. Behind this question nevertheless arises the question of the motivation of the researcher. The project of Charles and Perrine, this book makes clear, is at once a personal matter and a political act. The issues they are exploring are those of a sustainable future for our planet and our societies, the very challenges that scientists themselves need to assume and face. I share many of the ideas of Charles and Perrine. But my legitimacy to get involved as a scientist on their side is played out in the scientific nature of the problem, defining research questions, methodological choices that I could contribute. While it is necessary that the scientific methodology be scrutinized and critiqued, questions about the researcher's personal motivations aiming to undermine the scientific value of the research are void of scientific merit.

What justifies a special interest in the farm Bec? A pessimistic outlook might lead one to focus on the fact that Charles and Perrine invented nothing. Mollison and Holmgren gave us permaculture. Parisian gardeners in the nineteenth century, Eliot Coleman, John Jeavons, Pierre Rabhi, Philippe Desbrosses, and many others were there before them. Their inspirations are numerous, and they do not hide that fact, giving all their influences the same effort of reading and discovery. Their merit lies in the way they have assembled these references to promote the concept and practice of "ecological agriculture," which falls within the permaculture methodology in the way it was conceived, an overall construction of a cultivated and lived-in landscape, but also highlights professional intensive biological market gardening on a very small surface. This system was built gradually, and with trial and error they are finding their way. It will continue to evolve. This dynamic of apprenticeship in the service of a coherent and adaptive global project is at the core of La Ferme du Bec Hellouin, and it can and must serve as a source of inspiration to all those wishing to do the same elsewhere, with conviction that such a path is viable, even economically feasible given the study that we have completed. The reproducibility of this experiment lies not in the ability to do exactly the same thing elsewhere. It is the ability to understand the general logic and be inspired to read and act in other contexts, ecological, economic, human. To reach this understanding, this book, which is primarily a storybook, in my opinion, is the tool of choice. Apart from the recommendations and fact sheets, which are customary to agricultural extension work, it calls for more intelligence than compliance with established knowledge.

The mode of development experienced by agriculture in the last century was based on normative scientific approaches based on analytical determinations in the laboratory — even then, though, it was the field or experimental farm — where problems and solutions led to generic laws. These laws led to the universal definition and dissemination of technical and economic recommendations. The agroecological route, of which La Ferme du Bec Hellouin is an example, invites us to consider each agroeco-system as a complex and singular object, requiring unique solutions. This vision is the basis of my work as a researcher. Far from finding causalities and universal laws, my goal is to start from a thorough study of pioneering experiments to model the operation of small farms integrated into local markets, to identify rules that are adaptable for each situation, that guaran-tee sustainability in the short and medium term, and that are consistent with the issues and current and future agricultural constraints. I intend to contribute to building a holistic and interdisciplinary knowledge bank, which arms our ability to think and implement forms of agriculture that fall within the agroecological perspective, thinking in terms of immunity, metabolism, and functional integrity of agroecosystems; resilience and sustainability; social sense and strong sustainability. I share this objective with many other researchers around the world. We share the conviction that the future of agricultural science depends on our ability to provide avenues to disseminate pragmatic location-specific knowledge and scien-tific knowledge to design agroecosystems supported by a permanent dialogue between scientists and practitioners. It is to this dialogue that Charles and Perrine invited me. How could I refuse?

All these reasons are certainly excellent. But perhaps too cold and objec-tive to describe fully what animates me in this project. From a scientific perspective, I would like to say yes, of course. But I must be honest. My reasons are equally subjective. I like going to the Bec because it is rare and magical, owing much to its site within a small Normandy valley, no doubt, but the magic due mostly to the work and taste of Charles and Perrine.

What is striking is the care given to everything, to every detail. A care that meets a logical need: If there is anything that they have adopted from the Parisian market gardeners of the nineteenth century, it is that small and very well cared for is better than big and neglected. A principle that is undoubtedly an essential key to their success. But this care also meets an aesthetic logic, aimed at what many can feel, that we enjoy working more in this scenario and therefore work more efficiently.

At La Ferme du Bec Hellouin I have met groups of farmers, people wishing to create a farm, people who are merely curious. They always seem to have a sense of joy to be in this beautiful place, which makes them more attentive and confident in the projects that are shown to them. The farm is an agroecological utopia, and it is also a place where you feel good. This wellness suggests not only that the impossible is possible, but that we can take an active role in its realization. Those who live and work here are the measure of this beauty — gentle, open, curious, attentive to others. I do not come to the Bec only to speak in scientific terms about permaculture or reestablishing connections among societies, food, and nature. I go also to chat, and each of these chats makes me progress in my way as a scientist and a human being, more than serious and deep works and books.

— FRANÇOIS LÉGER[1]

EDITOR'S NOTE: As this translation goes to press, the study taking place at La Ferme du Bec Hellouin — to determine how much a market farmer working 1,000 square meters of land might earn — has drawn to a close. The final report (available at the farm web site, www.fermedubec.com) shows that during the last 12 months of the study, 1,600 hours were worked in the garden, and an additional 800 hours were consumed with related market-farming tasks, for a total of 2,400 work hours. The value of the product sold was 54,800 euros/$59,497. These results confirmed that it is indeed possible to earn one's living on a very small piece of land without use of mechanization using La Ferme du Bec Hellouin's method. They also confirmed that it is possible to engage an extra paid worker if the workload demands it.

ACKNOWLEDGMENTS

*L*a Ferme du Bec Hellouin is a human adventure to which many people have contributed. Our gratitude is deep for all of you who brought your stone to the building. You are really too numerous to mention here, but rest assured that we do not forget you!

To avoid getting anyone in trouble, it should be noted that the views expressed in these pages are our own and are not necessarily shared by persons and institutions listed below!

Our gratitude goes first to our parents and our families for their understanding and their unfailing support. You have done everything to help us, and we cannot thank you enough. Our thoughts go out especially to Marie-France, Perrine's mother, and François, Charles's father, who left us during these years. Thanks to our mini farmers, Lila, Rose, Shanti, and Fénoua, for your patience during the long evenings and weekends as we wrote this book, your tender support and helpful Internet research!

If the farm stands firm against the odds, it is thanks to a remarkable crew: Jean-Claude Bellencontre, Charles (Sacha) Guégan, Thomas Henriot, Édith Legay, Fabien Prud'homme, and not to forget Yohann Jourdan, Ludivine Ménard, or the amazing help of Jean-Pierre Bellencontre. Thank you to all of you for your commitment, your professionalism, your integrity, your humor, and your kindness. Working with you is a privilege every day.

The Bec Hellouin school of permaculture training is led by people we appreciate both for their human qualities and for the great skills that they bring to the work: thank you to Sylvain Barq, Pierre Boubarne, Armelle Devigon, Gérard Dufils, Marc Grollimund, and Agnès Sinaï, as well as occasional players. You teach us a lot and seeing you is always a joy! We thank the researchers that support on-farm studies: François Léger, through whom all began, Stéphane Bellon, Gauthier Chapelle, Cyril Girardin, Christine Aubry, Marc Dufumier, Pierre Stassart, Serge Valet, and students Justin Bourel, Morgane Goirand, Alexis de Liedekercke, and Kevin Morel.

A huge thank-you to the partners in our study, "Permacultural Organic Market-Farming and Economic Performance," whose true commitment has opened avenues to exploring tomorrow's agriculture: Fondation de France, Lemarchand Foundation for Balance Between People and the Earth, Léa Nature Foundation, Terra Symbiosis Foundation, Lunt Foundation, Fondation Pierre Rabhi, Picard Foundation.

Our gratitude is great for our friends who support the Institute Sylva, starting with its honorary president, Philippe Desbrosses, the current president, Lucile de Cossé Brissac, and Louis Albert de Broglie, Laurent de Chérisay, Sébastien Henry, and Alexandre Poussin. We also thank Jerome Henry and Crédit coopératif for their friendly complicity.

Volunteers and students who have completed internships at the farm are very numerous, and we assure each of them our gratitude. Thank you for the joint work and good times shared.

A special mention to Justin Bourel, who conducted a research project and provided corrections for this book! Thank you to Gauthier Chapelle for valuable remarks on biomimicry, our dear "coach" Sébastien Henry, and Sacha and Edith, who also helped edit this book.

Among the many people and friends who have supported and encouraged us throughout this adventure, we especially thank Claude Aubert, Huguette Autin, Peter Bal, Charles Barbot, Bernard Bertrand, Matthieu Blin, Antonin Bonnet, Hélène Bordeaux, Aurélie Bousselaire, Rachid Boutihane, Maëlys Bouttes, Robin Branchu, Sebastien Briant, Arnaud Brulaire, Alain Canet, Helen Square, Catherine Shalom, Yves Cochet, Eliot Coleman, Bruno Corroyer, Veronique Couvret, Arnaud Daguin, Benjamin Decooster, Guibert del Marmol, Christine Denis, Jacques Dereux, Sylvain Devaux, Cyril Dion, Jean-Baptiste Dumond, Violet and Claude Dumont, Alan Ferronière, Brother Raphael Flaujac, Jean-Martin Fortier, Jean-Yves Fromonot, Thierry and Pascale Glaizot, Gissinger Thierry, Bernard Gueral, Bruno Hermenault, Jean-Olivier Heron, Bertrand Hervieu, Jean-Paul and Blanche Hopsore, Mickaël Jammes, Sébastien Julliard, Aymeric Jung, Patricia Jung-Singh, Philippe Laborde, Sybille de Laboulaye, Frédéric Lamblin, Minnie and Romain Lassus, Tristan Lecomte, Vincent Legris, Florent Lemaire, Antoine Lemarchand, François and Françoise Lemarchand, Philip and Laura Lemarchand, Hélène Le Teno, Michael Lunt, Charlotte Mévius, Catherine Michaux, Camille Morel, Nguyen Ngoc Hanh,

Acknowledgments

Emmanuel Oblin, Serge and Carina Orru, Chantal Pessy, Ian Pinault, Nicolas and Marina Plowiecki, Marina Poiroux, Petra Popp, Mona Puill, Stephan Pierre Rabhi, Jean-François Renard, Maxime Rostolan, Emilia Wise Baptist Samson, Mickaël Santander, Claude Taleb, Claudius Thiriet, Sarah Valin, Marc Vandendrissche, Cyrille Varet, Françoise Vernet, Jean-Paul Vittecoq, Pauline Voghel, François Warlop. We also thank warmly the monks and nuns of Le Bec Hellouin. Without you, the farm would not be what it is! We owe you so much.

La Ferme du Bec Hellouin benefits from the support of the Upper Normandy region, the General Council of Eure, the Chamber of Agriculture of Eure, Pur Project, Le Bec Hellouin community, the GRAB of Haute-Normandie. We thank you for your confidence. Thank you to Welcome to the Farm, the tourism office of Brionne, and tourism committees of the department and the Haute-Normandy region.

Our gratitude is also extended to those working to develop permaculture in France and elsewhere: our "prof" Bernard Alonso, Jean-Philippe Beau-Douézy, Pascal Depienne, Franck Nathié, Frederick Proniewsky, Steve Read, Antoine Talin, Claire Uzan, Gildas Veret, Richard Wallner, teams from the People's University of Permaculture and Brin de Paille, and those we do not know. Moving Forward Together!

The original French version of this book is the fruit of friendly and patient perseverance of Jean-Paul Capitani and Françoise Nyssen, the Actes Sud leaders, supported by Clémence Beurton and Marie-Marie Andrascht, our editors, whom we thank warmly! Sharing a common passion with our publishers is precious. Thank you for the trust that you know how to create around you.

This American translation was possible thanks to Eliot Coleman, who introduced us to his wonderful publisher, Chelsea Green Publishing. Working with the Chelsea Green team made us forget the distance between France and the United States; we felt so close to them. Our special thanks goes to our translator, John F. Reynolds, who did a great job, and to our editor, Joni Praded. We also wish to thank Chelsea Green publisher, Margo Baldwin, and all the others who worked on the English edition of this book.

NOTES

EPIGRAPH

1. Théodore Monod, *Sortie de secours*, Seghers, 1991, p. 114.

INTRODUCTION

1. This number includes the hidden costs of our intensive agriculture, such as transportation, processing, packaging, storage, and distribution. Patrick Whitefield, *The Earth Care Manual*, Permanent Publications, 2004.
2. Quoted by Patrick Whitefield, ibid.
3. Energy is declining thanks to the progressive and inevitable diminuation of fossil resources.
4. See the excellent article "How Much Greenhouse Gas Was Emitted to Create Our Plate of Food?" from the website of Jean-Marc Jancovici, energy consulting engineer: www.manicore.com.
5. *The State of Food and Agriculture*, Food and Agriculture Organization, 2014, http://www.fao.org/3/a-i4036e.pdf.
6. This program and interim reports describing its progress are available online at www.fermedubec.com.
7. It is 1,000 square meters of cultivated space, to which should be added the circulation spaces, storage, et cetera. The farm can be of variable size.
8. The figures in the first year (2012–13) are presented in the Interim Report No. 2, available at www.fermedubec.com. The results of the following year largely confirm those of the first year.
9. Philippe Desbrosses, *Nous redeviendrons paysans*, Alpha, 5th edition, 2007.

ONE. PUPOLI'S CANOE

1. Jean Ziegler, *La Victoire des vaincus et résistance culturelle*, Éditions du Seuil, 1988, p. 12.
2. Robert Jaulin, *La Paix blanche. Introduction à l'ethnocide*, Éditions du Seuil, 1970, p. 19.
3. Quoted by Agnès Sinaï in "L'héritage aborigène aux sources de la permaculture," *La Revue durable* 50, October–December 2013, p. 19. (For an English version, please see *Permaculture: Principles and Pathways beyond Sustainability*, Holmgren Design Services, 2002, available through Chelsea Green Publishing.)
4. United Nations Organization for Agriculture and Development, www.fao.org/hunger/fr.
5. Department of Economic and Social Affairs, *World Economic and Social Survey 2011*, Overview, UN, 2011, p. 7.

6. According to Professor David Pimentel, from 1956 to 1996, 1.5 billion hectares (3.7 billion acres) of arable land were abandoned because of erosion. This is one-third of the arable surface of the planet. "Planète terre, planète désert?," May 3, 2007, www.liberterre.fr.

7. "The Area of Italy Lost Each Year (UN)," *La France Agricole*, October 22, 2010, www.lafranceagricole.fr.

8. This $20,000 is compared with the median income of households in the United States, which is approximately $50,000 (36,000 euros) per year. Selling the output of his garden can be an important source of additional income. US Census Bureau, www.census.gov.

9. Janine M. Benyus, Biomimétisme, Rue de l'Échiquier, coll. "Initial(e)s dd", 2011, citée sur www.mollat.com. (For an English version, see *Biomimicry: Innovation Inspired by Nature*, Harper Perennial, 2002.)

TWO. AROUND THE WORLD

1. Paul Virilio, *The Insecurity of Territory*, Stock, 1976.

2. Quoted in T. C. McLuhan, *Pieds nus sur la terre sacrée* [*Barefoot on the Sacred Earth*], Denoël, 1974, p. 8.

3. The adventures of *Fleur de Lampaul* were told in the form of fifteen books published by Gallimard Jeunesse and about a hundred television documentaries. Reports on the around-the-world tour can be viewed on or downloaded from the Internet under the title *Les Escales de Fleur de Lampaul* [*Ports of Call of Fleur de Lampaul*].

4. When we moved to the farm, I wrote a book summarizing what the Amerindian peoples taught me: *The Leaf Woman*, published by Albin Michel in 2007.

5. Hubert Reeves, *Travel Companions*, Seuil, coll. "Points," 2nd ed., 1998.

6. Francis Thompson, "The Mistress of Vision," *New Poems*, Copeland and Day, 1897.

7. For example: Paul Crutzen, Nobel Prize in Chemistry.

THREE. FROM DREAM TO REALITY

1. Quoted in T. C. McLuhan, *Pieds nus sur la terre sacrée.*

2. John Seymour, *The New Complete Book of Self-Sufficiency*, Dorling Kindersley, 3rd ed., 2003, translated into French under the title *Relive the Campaign*, De Boreas, coll. "Daily Life," 2007.

3. Perrine Hervé-Gruyer, *La relaxation en famille*, Presses de la Renaissance, 2008.

FOUR. THE AMAZON

1. Cited in Élisabeth Burgos, *Moi, Rigoberta Menchú*, Gallimard, 1983, p. 95 (rééd. 1999).

2. Albert Schweitzer, *À l'orée de la forêt vierge*, Albin Michel, 1995 (1re éd. 1929). (For an English version, see *The Primeval Forest*, Johns Hopkins University Press, 1998.)

FIVE. WE ARE WHAT WE EAT

1. Quoted in the journal *Terre du Ciel* 59.

2. See "Bébés, alerte aux pesticides!," December 2012, www.consoglobe.com.

3. American Cancer Society and International Agency for Research on Cancer (IARC), *Global Cancer Facts & Figures*, 3rd ed., February 4, 2015, www.cancer.org/acs/groups /content/@research/documents/document/acspc-044738.pdf.

SIX. DRAW ME A FARM

1. Quoted from their website Le Jardin en Mouvement: www.gillesclement.com.
2. See Lucien Pouëdras, *La Memorie des champs. La vie paysanne en Morbihan vers 1950*, Chasse-Maree/Armen, 1993.
3. A bit of semantics: In these pages, I have chosen to call an area cultivated naturally an agroecosystem, and an area cultivated intensively, more artificially, an agrosystem.

SEVEN. NEW FARMERS

1. Jean-Philippe Barde and Christian Garnier, *L'Environnement sans Frontieres*, Seghers, 1971.
2. Quoted in Monod, *Sortie de secours*.

EIGHT. DISCOVERING PERMACULTURE

1. Bill Mollison and David Holmgren, *Permaculture One: A Perennial Agriculture for Human Settlements*, International Tree Crop Institute, 1981.
2. Sociocracy is a mode of decision making and governance that allows an organization, whatever its size — from a family to a country — to behave like a living organism, to self-organize (Wikipedia).
3. "A local exchange system is an exchange of products or services that are within a closed group (usually associations)" (Wikipedia).
4. See Whitefield, *Earth Care Manual*, and *How to Grow More Vegetables than You Ever Thought Possible on Less Land than You Can Imagine*, Ten Speed Press, 8th ed., 2012.
5. In gardening, a cultivated strip of land is called a bed.
6. Richard Wallner published *Un Manuel de la culture sur butte* (Rustica, 2013). He also translates and distributes various books and DVDS on permaculture.
7. The book you have in your hands discusses the concepts of permaculture and bio-inspired agriculture. We plan to write an illustrated practical guide that will describe, in detail, the applications of all topics discussed in these pages, to be published by Actes Sud.
8. The certified permaculture course (CCP), or Permaculture Design Course (PDC) in English, constitutes the first level of training in permaculture. It was established by Bill Mollison and remains the internationally recognized standard format. Its duration is at least seventy-two hours, and usually eleven days of instruction. The CCP provides an introduction to the major themes of permaculture. This first level of study may be extended by a personalized program.
9. Surely it would be better to buy manure from an organic farm, or to produce it ourselves, which we do in small quantities with our animals. But the nearest organic farm is tens of kilometers away. European legislation on organic farming allows the use of non-organic manure, provided it does not come from factory farms. Still, we

hope that in the near future, the development of organic farming will enable growers to dispose of organic manure locally.

10. See www.agroforestry.co.uk.

11. Chipping wood is a technique developed by Canadian researchers to make use of small sections of wood branches (less than 5 centimeters/2 inches). Live young twigs are a concentration of minerals, proteins, hormones, and bio-catalysts. Dispersed as mulch on the ground or incorporated, wood chips improve depleted soil. An agro-sylvo-pastoral system includes crops, trees, and animals.

12. His book *Sepp Holzer's Permaculture: A Practical Guide to Small-Scale, Integrative Farming and Gardening*, published by Chelesa Green in 2011, has not finished making us dream!

13. Emmanuel Oblin sells the best tools he could find in his research. His website hosts numerous videos that allow you to learn the art of mowing and threshing and the sharpening of the blade: www.comptoirdelafaux.com.

NINE. BIOINTENSIVE MICROAGRICULTURE

1. John Jeavons, *How to Grow More Vegetables*, Ten Speed Press, 2012. Editor's note: The data in this chapter are from an earlier 2006 edition. The 2012 edition contains many updates and new data.

2. A broadfork is a type of fork with two handles and long tines. Farmers use the tool to decompact the soil.

3. Whitefield, *Earth Care Manual*.

4. Jeavons, *How to Grow More Vegetables*, Ten Speed Press, 2012. Editor's note: See note 1.

5. Quoted by John Jeavons in ibid., p. xviii.

6. Ibid., p. xii.

7. Ibid., p. vi.

8. Ibid., p. xii.

9. "The yields of wheat and maize not progressing," *Agreste Primeur* 210 (official document of the Ministry of Agriculture), May 2008. To learn more about gardening wheat, read Joseph Pousset, *Treaty of Agroecology*, agricultural France, coll. "Agriproduction," 2nd ed., 2012.

10. See http://www.growbiointensive.org/grow_main.html.

11. François Couplan, *Nutritional Guide of Wild and Cultivated Plants*, Delachaux and Niestlé, 1998.

TEN. ELIOT COLEMAN

1. Conversation with the author.

2. Eliot Coleman, *The Winter Harvest Handbook*, Chelsea Green Publishing, 2009.

3. Abraham Lincoln, quoted by Jeavons, *How to Grow More Vegetables*, Ten Speed Press, 2012. It is worth noting that in the original quote, Lincoln went on to say, "No community whose every member possesses this art, can ever be the victim of oppression of any of its forms. Such community will be alike independent of crowned-kings, money-kings, and land-kings."

4. Eliot Coleman, *Winter Harvest Handbook*, Chelsea Green Publishing, 2009.

5. Eliot Coleman, *Four Season Harvest*, Chelsea Green Publishing, 1999, and *The New Organic Grower*, Chelsea Green Publishing, 1995.

6. Coleman still has two tractors and a large professional rototiller, but they are used only very occasionally.

7. The company's website, www.johnnyseeds.com, is interesting, illustrated with numerous instructional videos. You can order Coleman tools there.

8. Jean-Martin Fortier, *The Market Gardener*, New Society Publishers, 2014.

ELEVEN. THE PARISIAN MARKET GARDENERS OF THE NINETEENTH CENTURY

1. I. Ponce, *La Culture maraîchère pratique des environs de Paris*, Librairie agricole de la Maison rustique, 1869.

2. J. Cure, *Ma practique de la culture maraîchère ordinaire et forceè*, 1918.

3. Jeavons, *How to Grow More Vegetables*, Ten Speed Press, 2012.

4. For those who want to know more about Kropotkin, a prominent anarchist and scholar who in the 1800s wrote on the role of mutual aid in evolution, see http://dwardmac.pitzer.edu/Anarchist_Archives/kropotkin/Kropotkinarchive.html. For an article briefly describing his work, see Lee Alan Dugatkin, "The Russian Anarchist Prince Who Challenged Evolution," *Slate*, October 30, 2012, accessed September 21, 2015, http://www.slate.com/articles/health_and_science/human_evolution/2012/10/evolution_of_cooperation_russian_anarchist_prince_peter_kropotkin_and_the.html.

5. See Eliot Coleman, *Winter Harvest Handbook*, Chelsea Green Publishing, 2009.

6. J. G. Moreau and J. J. Daverne, *Manuel pratique de la culture maraîchère à Paris*, 1845, p. 83.

7. Readers interested in the question may consult these books in the documentary collection of the Bec Hellouin permaculture school, www.ecoledepermaculture.org.

8. Moreau and Daverne, *Manuel pratique de la culture maraîchère à Paris*, p. 7.

9. Ibid., p. 15.

10. Ibid., p. 84.

11. A. Dumas, *La Culture maraîchère: traité pratique pour le Midi, le Centre de la France et pour l'Algérie*, J. Rothschild, 4th ed, 1880.

12. Ponce, *La Culture maraîchère pratique des environs de Paris*.

13. The European regulations for organic farming set the acceptable limit of nitrogen per hectare (2.5 acres) at 170 kilograms (375 pounds) per year.

14. How can we, in the future, make these thermal blankets and plastic films without oil? This question is worth pursuing!

15. In comparison, recent technical and economic studies on organic market gardening in France give the figure as an average value of 1.2 rotations per year.

TWELVE. EXOTIC INFLUENCE

1. Azby Brown, *Just Enough: Lessons in Living Green from Traditional Japan*, Tuttle Publishing, 2012.

2. Dena Merriam, *The Message in a Seed: Guidelines for Peaceful Living*, Shumei International, 2007, p. 8.
3. In France we call this kind of organization an AMAP (Association pour le maintien d'une agriculture paysanne). I like the meaning of the North American name, community-supported agriculture.
4. Masanobu Fukuoka, *The One-Straw Revolution*, Guy Trédaniel publisher, 2005. (For the English edition, please see *The One-Straw Revolution*, 1978, Rodale.)
5. For more on this subject, see Laurence Green, *A Pilot Study Comparing Gaseous Emissions Associated with Organic Waste Treated with and Without Bokashi Fermentation*, 2009, www.bokashicycle.com.
6. See the website of Purin d'Orties and Co. for this article: "Les micro-organismes efficaces: eM," www.purindortie-bretagne.com.
7. See "The Real Dirt on Rainforest Fertility," *Science* 297, August 9, 2002, p. 921.
8. www.lafabriculture.fr.

THIRTEEN. GENESIS OF A METHOD

1. Website: www.formationsbio.com.
2. It should be noted that the subsidies received to construct the ecocenter — subsidies motivated by the coherence of this project with the regional policies on tourism and training — should not mask the lack of support that generally accompanies the creation of an organic farm, in the current context. In our case, outside of the assistance for training young farmers, the only subsidy received when we created our farm was the "boost aid" from the Haute-Normandie region, an amount of 3,000 euros. It seems illusory to rely too much on public support when one aspires to embark on such an adventure.
3. Safer is the organization for land development and rural settlement.
4. La Ferme du Bec Hellouin method is described at www.fermedubec.com.

FOURTEEN. LAUNCH OF A RESEARCH PROGRAM

1. Quoted in Jean-Paul Besset and René Dumont, *Une vie saisie par l'écologie*, Stock, 1992.
2. The initial project and interim reports are available on the www.fermedubec.com site.
3. Over the next two years, other organizations have launched support for the project: the Terra Symbiosis Foundation, the Lunt Foundation, Fondation Pierre Rabhi, the Picard Foundation.
4. François Léger, UMR SAD-APT Paris, AgroParisTech, agroecology; Christine Aubry, UMR SAD-APT Paris, INRA SAD, agronomy; Stéphane Bellon, UR Ecodevelopment Avignon INRA SAD, agro-ecology; Marc Dufumier, UFR Ecodevelopment Avignon, INRA SAD, agroecology; Mark Dufumier, UFR Agriculture comparison and development, AgroParisTech; Philippe Desbrosses, Sainte Marthe farm, agronomy; Pierre Stassart, University of Liège, sociology; Gauthier Chapelle, Greenloop, biomimicry; Serge Valet, agronomy.
5. The microfarm La Bourdaisière was launched at the initiative of Louis-Albert de Broglie, the dynamic website owner, and his collaborator Maxime Rostolan. More info at www.fermesdavenir.org.

6. Gauthier Chapelle founded Biomimicry-Europa and Greenloop, based in Brussels. Biomimicry draws inspiration from nature (its forms, materials, and mechanisms) to achieve or optimize human creations: imitating the operation of a whale's heart to create new pacemakers in the form of a patch; seeking inspiration from the structure of silk to create materials more resistant than steel; copying the design of a mosquito's proboscis to create painless needles.

7. The chaffinch (*Fringilla coelebs*), European greenfinch (*Chloris chloris*), European goldfinch (*Carduelis carduelis*), common linnet (*Linaria cannabina*), European serin (*Serinus serinus*), Eurasian bullfinch (*Pyrrhula pyrrhula*), and hawfinch (*Coccothraustes coccothraustes*).

8. A decline of, respectively, 70 and 60 percent in France over the last twenty years.

9. *Trichodes alvearius.* For a completely anecdotal comparison: In two days in June, I have seen at least two dozen — as many as all of the Belgian entries combined between June 1 and July 15 — on the site: www.observations.be.

10. A very involved permaculturist, Jean-Philippe Beau-Douézy is the founder of the ecocenter Le Bouchot: www.lebouchot.net.

FIFTEEN. THE FOREST GARDEN

1. Henry David Thoreau, from his essay "Walking," available from *The Atlantic* at http://www.theatlantic.com/magazine/archive/1862/06/walking/304674.

2. The forest garden is called various names by different authors. My preferred term is *forest garden*, which best describes the priority given to trees.

3. Patrick Whitefield, *How to Make a Forest Garden*, Permanent Publications, UK, 2002, page xv.

4. On this subject, there's an excellent book by Matthew Calame, *La Tourmente alimentaire*, Charles Léopold Mayer, 2008.

5. See the websites www.agroforestry.co.uk and www.pfaf.org.

6. See the website www.foretscomestibles.com.

7. Benjamin Hennot, *La Jungle étroite*, 2013.

8. See Robert Hart, *Forest Gardening*, Chelsea Green Publishing, 2nd ed., 1996.

9. Ibid., p. xvi.

10. In tropical countries.

SIXTEEN. AGRICULTURE OF THE SUN

1. Antonio Gramsci, *Cahiers de prison*, Gallimard, 1996, t. I, Book 3, p. 283. (For an English version, see *The Prison Notebooks*, Columbia University Press, 2011.)

2. Interview with Clemens Heller in Paris in 1949, famously reported by Nancy Lutkehaus in *Margaret Mead: The Making of an American Icon*, Princeton University Press, 2008.

3. All data are from the WWF, *Living Planet Report*, published in 2012, p. 36.

4. An ecological footprint is the area required to meet the resource needs of a person and absorb his or her waste.

5. Oxfam, "Ending Extreme Inequalities," January 20, 2014. Available at www.oxfam.org.

6. A survey of the Global Ecovillage Network has calculated that at Findhorn, an ecovillage located in Scotland, an inhabitant has an average ecological footprint two times lighter than that of a typical inhabitant of the United Kingdom — which is still too much!

7. See *Farmers' Handbook*, 2010, www.permaculture.org.au.

8. See *Libro: Permacultura criolla*, 2010, www.permacultura-es.org.

9. Food portion of the ecological footprint of Europeans: 30 percent according to *Our Ecological Footprint*; 20 percent according Enviro-meter; studies cited in Whitefield, *Earth Care Manual*. Share of food's impact on the carbon footprint of the French: 22 percent according to a presentation by Alain Grandjean in 2013; 31.6 percent according to Jean-Marc Jancovici in the article "How Much Greenhouse Gases in Our Plate?" published in January 2010 on his blog (www.manicore.com). The latter figure includes the energy used by trucks carrying our food, by supermarkets, by us as we make trips to and from the supermarket, and so on. In fact all these measurements depend on the embodied energy that is included or excluded when we talk about food.

10. Summary of various studies carried out by Datamatch: "World Energy Outlook," International Energy Agency, 2013; The Shift Project, www.tsp-data-portal.org; Jean-Marc Jancovici, www.manicore.com; US Energy Information Administration, www.eia.gov; among others. The summary was published in *Paris Match* 3369, December 12–18, 2013.

11. Ibid.

12. Ibid.

13. Ibid.

14. Benyus, *Biomimicry*, p. 38.

15. Ibid, p. 39.

16. Whitefield, *Earth Care Manual*. Also on the website of IBM (www.ibm com.): "An optimized system for a smarter planet."

17. Study quoted in Nicolas Lampkin, *Organic Farming*, Farming Press, 2003.

18. Sébastien Debande, *La Permaculture, un intérêt économique*, 2010 (downloadable memory http://fr.scribd.com/doc/32224788/ La-PermacuLture-un-interet-economique).

19. See Jean Berthier, *Les Routes, les ponts et les parcs de stationnement*, Techniques de l'ingénieur, August 10, 2010, "Évolution du parc automobile français" (table), www.techniques-ingenieur.fr; Institut national d'études démographiques, "France-Allemagne: histoire d'un chassé-croisé démographique," *Population & Sociétés* 487, March 2012.

20. See Jean-Noël Jeanneney and Jeanne Guérout, *Jours de guerre*, Les Arènes, 2013.

21. David Holmgren, *Permaculture: Principles and Pathways Beyond Sustainability*, Holmgren Design Services, 2002.

22. Tropical forests annually produce 2,200 grams (4.9 pounds) of dry organic matter per square meter on average; deciduous temperate forests, 1,200 grams (2.6 pounds); temperate grasslands, 600 grams (1.3 pounds); croplands in all climates, 650 grams (1.4 pounds). Whitefield, *Earth Care Manual*, p. 22.

23. This is the approach of the Hummingbirds Movement, founded by Pierre Rabhi and Cyril Dion, actively disseminating solutions "that work": www.colibris-lemouvement.org.

24. According to *Sources* 26, p. 86.

SEVENTEEN. WORKING BY HAND

1. Holmgren, *Permaculture.*
2. The energy in fossil fuels comes from the sun. Millions of years ago it was captured by carbon-based life-forms — plants and animals — and concentrated over time, as they decayed, into the fossil fuel we extract today.
3. See Manual Rovillée, "Le ver de terre, star du sol," www.cnrs.fr.
4. See the work of Marcel Bouché, researcher and reputed géodrilologue, to whom we owe the first mapping of earthworms in France, and his book *Des vers de terre et des hommes: découvrir nos écosystèmes fonctionnant à l'énergie solaire*, Actes Sud, 2014.
5. See Patricia Hanssens, "La Fraternité ouvrière ou les Tropiques à Mouscron," *Le 23* 204 (the magazine of Maison Régionale de l'Environnement et des Solidarités, a regional network of environmental, solidarity, and human rights organizations), Summer 2011, p. 14.
6. Communication with Alain Canet, president of the French Association of Agroforestry.
7. "The Area of Italy Lost Each Year (UN)," *La France Agricole.*
8. *La Revue durable* 50.
9. The creation of soil differs on a case-by-case basis, though new soil is created on the order of 1 centimeter (0.4 inch) per century; two thousand years is sometimes necessary to create 10cm humus! See also "Disparition des terres agricoles en France," www.planetoscope.com.
10. An amateur gardener who performs a rotation of crops each year in the family garden does not have the same fertility needs as the market gardener who performs three to eight rotations per year, as is the case at La Ferme du Bec Hellouin.
11. Currently, almost all farmers subsidized by Europe destroy soils. In some contexts, the more they consume, the more you pay! See "How a False Solution to Climate Change Is Damaging the Natural World" on the *Guardian* website, www.theguardian.com.
12. As this translation goes to press, the study has just finished and the final report is on our website www.fermedubec.com. During the last 12 months of the survey, 1600 hours were worked in the garden, and an additional 800 hours were consumed with related market-farming tasks, for a total of 2,400 work hours. The value of the product sold was 54,800 euros ($59,497). The data show that a market gardener working entirely by hand on 1,000 square meters using our method makes as much money per hour as a market gardener with a tractor — as John Jeavons said in the seventies.
13. On holistic grassland and cattle management, see the interesting work of Alan Savory (http://savory.global) and Frédéric Thomas.
14. Judith Soule and Jon Piper, *Farming in Nature's Image*, Island Press, 1992.
15. A food web includes all food chains in an ecosystem.
16. I have seen, in Kerala, India, a dramatic shift from traditional fishing, providing the local people with animal protein at low prices, to industrial fishing that transforms the

fish to bonemeal and exports it to rich countries, depriving the population of this valuable resource and eliminating many jobs.

17. Olivier Blond, "Manger tue," www.goodplanet.info, February 13, 2014.
18. See Valérie Orsini, "Le coût de l'obésité aux États-Unis: 500 milliards de dollars d'ici 2030," www.atlantico.fr, May 24, 2012.
19. See Muriel Levet, *Ces peuples sans maladies: leurs secrets de longévité*, Trajectoire 2007.
20. See Benyus, *Biomimicry*, p. 46.
21. See Whitefield, *Earth Care Manual*, p. 18.
22. Ibid.
23. Pousset, *Treaty of Agroecology.*
24. Wes Jackson and The Land Institute in the United States.
25. See www.farinesdemeule.com.
26. On these issues, see the latest books by Marc Dufumier: *Cinquante idées reçues sur l'agri- culture et l'alimentation*, Allary Publishing, 2014, and *Famine au Sud, malbouffe au Nord: comment le bio peut nous sauver*, Nile, 2012.
27. Jacques Légeret, *L'Énigme amish: vivre au xxie siècle comme au xviie*, Labor et Fides, 2000, p. 148.
28. Ibid., p. 149.
29. John A. Hostetler, *Amish Roots*, John Hopkins University Press, 1989.
30. Légeret, *L'Énigme amish*, p. 154.

EIGHTEEN. TO BE SMALL

1. Robert Blondin, *Le Bonheur possible*, Les Éditions de l'Homme, 1983, p. 59.
2. See Oxfam, https://www.oxfam.org/en/tags/small-scale-farming.
3. See Alimenterre, "Les paradoxes de la faim," www.alimenterre.org.
4. See Whitefield, *Earth Care Manual*, p. 31.
5. See Debande, *La Permaculture, un intérêt économique.*
6. The production recorded during 2013 as part of the study at La Ferme du Bec Hellouin represents seventy-eight weekly baskets at 10 euros ($11.30), fifty weeks per year.
7. See Debande, *La Permaculture, un intérêt économique.*
8. Professional market gardeners believe that if we deduct for paths and tractor wheels, the actual area is about 7,000 square meters.

NINETEEN. MICROFARMS

1. From E. F. Schumacher's classic collection of essays, *Small Is Beautiful: A Study of Economics as If People Mattered*, originally published in 1973 by Blond and Briggs.
2. Quoted in Besset and Dumont, *Une vie saisie par l'écologie*, p. 172.
3. The alter-globalization movement, also called the global justice movement, advocates alternative forms of globalization — such as global cooperation and knowledge sharing — but opposes economic globalization because of its negative consequences.
4. *Bijogos, les Grands Hommes de l'archipel*, Gallimard Jeunesse, coll. "*Fleur de Lampaul*," 1993.
5. We strongly recommend reading *La Revue durable*, with its particularly well-documented environmental themes.

6. The average ecological footprint of a US citizen is 7.19 global hectares (gha), against 0.66 gha for Bangladeshis. WWF, *Living Planet Report*, pp. 142, 144.

7. On this subject, read the interesting report by Pablo Servigne: *Nourrir l'Europe en temps de crise*, published in 2013, citing La Ferme du Bec Hellouin: http://www.greens-efa.eu.

8. Such projects already exist in various forms in different parts of France.

9. See ben-law.co.uk.

10. See Jonathon Porritt, *Save the Earth*, DK, 1992.

11. A vast amount of literature on agroforestry can be found with the French Association of Agroforestry.

12. From the report *Nourrir l'Europe en temps de crise*, Pablo Servigne speaks about 117 million Europeans to form a generation (p. 35).

13. www.la-fee.org.

14. "In soil we trust" is a riff on the phrase found on US currency, "In God we trust." For more information: www.slowmoney.org and Woody Tasch, *Slow Money: Inquiries into the Nature of Slow Money*, Chelsea Green Publishing, 2010.

TWENTY. MICROAGRICULTURE, SOCIETY, PLANET

1. Robert Blondin, *Le Bonheur possible*, p. 10.

2. Source: Insee, the National Institute of Statistics and Economic Studies.

3. www.agreste.agriculture.gouv.fr, *Agreste Primeur* 266, September 2011, the figure for 2010, and the document *L'Agriculture française depuis cinquante ans*.

4. "Panorama de l'agriculture," www.lafranceagricole.fr.

5. www.insee.fr, table "Population active et taux d'activité selon la nationalité le sexe et l'âge en 2012"; "Demandeurs d'emploi inscrits et offres collectées par Pôle emploi en décembre 2013," www. travail-emploi.gouv.fr.

6. "Demandeurs d'emploi inscrits et offres collectées par Pôle emploi en décembre 2013," www. travail-emploi.gouv.fr.

7. ww.insee.fr, table "Population, superficie et densité des principaux pays du monde en 2013."

8. www.insee.fr, table "Exploitations agricoles selon la superficie agricole utilisée en 2010."

9. See Institut national de l'information géographique et forestière, "La surface forestière en France métropolitaine," www.inventaire-forestier.ign.fr.

10. See http://www.unep.org/Documents.Multilingual/Default.asp?ArticleID=5417&DocumentID=485&l=en.

11. France is an exporter of agricultural products. An additional area may be allocated to crops for export, however, if this is relevant in the biosphere and to the peoples of the south — relevance that does appear to be the case.

12. The Loess Plateau, in China, the size of Belgium, has been reforested after being desertified by thousands of years of subsistence farming. It was necessary to convince farmers to plant trees "that they do not eat," but the operation, after ten years, was a great success. See the short film *Hope in a Changing Climate* by John D. Liu, https://www.youtube.com/watch?v=bLdNhZ6kAzo.

13. Quoted by Monod, *Sortie de secours*.
14. Quoted in Porritt, *Save the Earth*, DK, 1992. p. 115.
15. Source Insee.
16. See *Carbone 4*, a presentation of Alain Grandjean, www.carbone4.com, 2013.
17. Pierre Rabhi, "In Praise of the Creative Genius of the Civil Society," *Actes Sud*, coll. "Domaine possible," 2011.

TWENTY-ONE. THE EARTH IS AN ADVENTURE

1. Quoted by Nelson Mandela in his inaugural address in 1994 and originating in Marianne Williamson, *A Return to Love*, HarperCollins, 1992.
2. Sébastien Henry has written two marvelous books: *Quand les décideurs s'inspirent des moines*, Dunod, 2012, and *Ces décideurs qui méditent et s'engagent*, Dunod, 2014.

TWENTY-TWO. BIO-ABUNDANCE

1. André Breton, *Arcane 17*.
2. Quoted in Théodore Monod, *Et si l'aventure humaine devait échouer*, Grasset, 2004.
3. All those interested in this issue will benefit from reading the work of James Lovelock, *Gaia: How to Cure a Sick Earth*, Robert Laffont, 1992.
4. From the article by René Passet, "Un développement contre nature," *Calypso Log* 115, September 1992.

AFTERWORD

1. Agricultural engineer, doctor of ecology. Research professor at AgroParisTech, member of UMR SAD-APT Science for Action and Sustainable Development: Activities, Products, Territories (INRA AgroParisTech). Associate researcher at UMR Eco-Anthropology and Ethnobiology (CNRS-MNHN). Chairman of the Scientific and Technical Council of the national bodies to rural and agricultural vocation. Member of the Scientific Council of the Coastal Conservancy. Director of UMR SAD-APT 2006–13.

RESOURCES

*A*t La Ferme du Bec Hellouin, there are books everywhere. Here is a small selection of those that inspired us and are available in English. You will find hundreds of other books and resources — available in French, English, or both — in the bibliography, indexed by subject matter, archived on our website for the farm's permaculture school. (See www.fermedubec.com.) There, in addition to the categories listed below, you'll find resources on beekeeping, crop mixing, husbandry, fertility, biodynamics, seeds, tools, trees, aromatic and medicinal plants, flowers, bio-indicator plants, wild plants, butterflies, natural landscapes, and more. You will also find the scientific reports on the studies conducted at the farm and our films.

FOOD PRODUCTION, PERMACULTURE DESIGN, AND AGROECOLOGY

Aranya. *Permaculture Design: A Step-by-Step Guide*. Hampshire, England: Permanent Publications, 2012.

Benyus, Janine M. *Biomimicry: Innovation Inspired by Nature*. New York: William Morrow & Co., 1997.

Crawford, Martin. *Creating a Forest Garden: Working with Nature to Grow Edible Crops*. Totnes, England: Green Books, 2010.

Falk, Ben. *The Resilient Farm and Homestead: An Innovative Permaculture and Whole Systems Design Approach*. White River Jct, VT: Chelsea Green Publishing, 2013.

Fukuoka, Masanobu. *The One-Straw Revolution: An Introduction to Natural Farming*. Translated by Chris Pearce, Tsune Kurosawa, and Larry Korn. Emmaus, MA: Rodale Press, 1978.

Goldring, Andrew. *Permaculture Teachers' Guide*. London: Permaculture Association/ WWF-UK, 2000.

Hemenway, Toby. *Gaia's Garden: A Guide to Home-Scale Permaculture*. White Rivt Jct, VT: Chelsea Green Publishing, 2000.

Hart, Robert. *Forest Gardening*. 2nd ed. White River Jct, VT: Chelsea Green Publishing, 1996.

Holmgren, David. *Permaculture: Principles & Pathways Beyond Sustainability*. Hepburn, Victoria, Canada: Holmgren Design Services, 2002.

Holmgren, David. *Permaculture Pioneers: Stories from the New Frontier*. Edited by Kerry Dawborn, and Caroline Smith. Hepburn, Victoria, Canada: Holmgren Design Services, 2011.

Holzer, Sepp. *Sepp Holzer's Permaculture: A Practical Guide to Small-Scale Integrative Farming and Gardening*. White River Jct, VT: Chelsea Green Publishing, 2010.

Jacke, Dave, and Eric Toensmeier. *Edible Forest Gardens*, Vol. 1, *Design and Practice*; Vol.2: *Vision and Theory*. White River Jct, VT: Chelsea Green Publishing, 2005.

Mollison, Bill, and David Holmgren. *Permaculture One: A Perennial Agriculture for Human Settlements*. Tyalgum, New South Wales, Australia: Transworld, 1987.

Mollison, Bill, and Reny Mia Slay. *Introduction to Permaculture*. Tyalgum, New South Wales. Australia: Ten Speed Press; Revised edition, 1997.

Mollison, Bill. *Permaculture Two: Practical Design for Town and Country in Permanent Agriculture*. Stanley, Tasmania, Australia: Tagari Publications, 1979.

Law, Ben. *The Woodland Year*. Hampshire, England: Permanent Publications, 2008.

Law, Ben. *The Woodland Way: A Permaculture Approach to Sustainable Woodland Management*. 2nd ed. Hampshire, England: Permanent Publications, 2013.

Whitefield, Patrick. *Permaculture in a Nutshell*. Clanfield, England: Permanent Publications, 2011.

Whitefield, Patrick. *The Earth Care Manual: A Permaculture Handbook for Britain and Other Temperate Climates*. Portsmouth, England: Permanent Publications, 2004.

Whitefield, Patrick. *How to Make a Forest Garden*. 3rd ed. Clanfield, England: Permanent Publications, 2002.

SOIL AND WATER MANAGEMENT

Lewis, Wayne, and Jeff Lowenfels. *Teaming with Microbes: The Organic Gardener's Guide to the Soil Food Web*. Portland, OR: Timber Press, 2010.

Holzer, Sepp. *Desert or Paradise: Restoring Endangered Landscapes Using Water Management, Including Lake and Pond Construction*. White River Jct, VT: Chelsea Green Publishing, 2012.

Lancaster, Brad. *Rainwater Harvesting*. 3 volumes. Tucson, AZ: Rainsource Press, 2007; White River Jct, VT: Chelsea Green Publishing, 2013.

Yeomans, P.A. *Water for Every Farm: Yoemans Keyline Plan*. 4th ed. Southport, Queensland, Australia: Keyline Design, 2008.

Caddy, Eileen and Peter. *The Findhorn Garden Story*. 3rd ed. Findhorn, Scotland: Findhorn Press, 2008.

Wright, Machaelle Small. *The Perelandra Garden Workbook: A Complete Guide to Gardening with Nature Intelligences*. 2nd ed. Jeffersonton, VA: Perelandra, 1993.

HOME-SCALE GROWING
AND FARM-LIFE GUIDES

Dowding, Charles. *Salad Leaves for All Seasons: Organic Growing from Pot to Plot*. Totnes, England: Green Books, 2008.

Fern, Ken. *Plants for a Future: Edible & Useful Plants for a Healthier World*, Permanent Publications, 2nd ed., 2009.

Jeavons, John. *How to Grow More Vegetables (And Fruits, Nuts, Berries, Grains, and Other Crops) Than You Ever Thought Possible on Less Land Than You Can Imagine*. 8th ed. Berkeley: Ten Speed Press, 2012.

Lazor, Jack. *The Organic Grain Grower: Small-Scale, Holistic Grain Production for the Home and Market Producer*. White River Jct, VT: Chelsea Green Publishing, 2013.

Ruppenthal, R J. *Fresh Food from Small Spaces: The Square-Inch Gardener's Guide to Year-Round Growing, Fermenting, and Sprouting*. White River Jct: Chelsea Green Publishing, 2008.

Woram, Catherine, and Martyn Cox. *Gardening with Kids*. New York: Ryland Peters & Small, 2008.

MARKET FARMING

Coleman, Eliot. *Four-Season Harvest: Organic Vegetables from Your Home Garden All Year Long*. 2nd ed. White River Jct, VT: Chelsea Green, 1999.

Coleman, Eliot. *The New Organic Grower: A Master's Manual of Tools and Techniques for the Home and Market Gardener*. White River Jct, VT: Chelsea Green Publishing, 2nd ed, 1995.

Coleman, Eliot. *The Winter Harvest Handbook: Year Round Vegetable Production Using Deep Organic Techniques and Unheated Greenhouses*. White River Jct, VT: Chelsea Green Publishing, 2009.

Martin-Fortier, Jean. *The Market Gardener: A Successful Grower's Handbook for Small-Scale Organic Farming*. Gabriola Island, Bristish Columbia, Canada: New Society Publishers, 2014.

FOOD PROCESSING

Crawford, Martin, and Caroline Aitken. *Food from Your Forest Garden: How to Harvest, Cook, and Preserve Your Forest Garden Produce*. Totnes, England: Green Books, 2013.

MUSHROOMS

Stamets, Paul. *Growing Gourmet and Medicinal Mushrooms*. 3rd ed. Berkeley, CA: Ten Speed Press, 2000.

BIRDS

Burton, Robert. *Bird Behavior*. 1st American ed. New York: Knopf, 1985.

ART OF LIVING

Brown, Azby. *Just Enough: Lessons in Living Green from Traditional Japan*. North Clarendon, VT: Tuttle Publishing, 2013.

Hopkins, Rob. *The Transition Handbook: From Oil Dependency to Local Resilience*. UIT Cambridge Ltd., 2014.

Lovelock, James. *Gaia: A New Look at Life on Earth*. Subsequent ed. Oxford, England: Oxford Paperbacks, 2000.

Merriam, Dena. *The Message in a Seed: Guidelines for Peaceful Living*. Shumei International, 2007.

Seymour, John. *The New Complete Book of Self-Sufficiency: The Classic Guide for Realists and Dreamers*. London: Dorling Kindersley, 2009.

HABITAT, CONSTRUCTION, AND ARTISAN CRAFT

Dawborn, Kerry, and Caroline Smith. *Permaculture Pioneers: Stories from the New Frontier*. Hepburn, Victoria, Canada: Mellidora Publishing, 2011.

Erdoes, Richard. *Crying for a Dream: The World through Native American Eyes*. 2nd ed. Rochester, VT: Bear & Company, 2001.

Mills, Edward, and Rebecca Oaks. *Greenwood Crafts: A Comprehensive Guide*. Ramsbury, England: Crowood Press, 2012.

Seymour, John. *The Forgotten Arts and Crafts*. 1st America ed., annotated. New York: Dorling Kindersley, 2001.

Snell, Clarke, and Timothy L. Callahan. *Building Green: A Complete How-To Guide to Alternative Building Methods: Earth Plaster, Straw Bale, Cordwood, Cob, Living Roofs*. New York: Lark Books, 2005.

INDEX

Note: *ci* refers to the Color Insert

Index

Morris, Desmond, 202

mulch. *See also* compost
  benefits of, 92
  ferns as, 184
  grain growing and, 161, 185
  nettles as, 57–58
  raised beds and, 53, 80
  slugs and, 64, 101
  soil-building and, 149
  terraced gardens and, 60–61
Müller, Hans, 99
mycorrhizal fungi, 94, 149

Nathié, Franck, 133
National Institute for Agricultural
  Research (INRA), 4, 117
nature, 1, 9–10, 201. *See also* wildlife
Nature et Découvertes, 111
Nearing, Helen and Scott, 74
Nepal, 139
nettles, 57–58, 81
The New Organic Grower (Coleman), 75
nonaction, 99–100
no-till agriculture, 147–152, 167
Nous redeviendrons paysans (Desbrosses),
  110
nursery production, 179
Nyssen, Françoise, 73

Oblin, Emmanuel, 61
oil, 3, 70, 139–143
Okada, Mokichi, 98–99
The One-Straw Revolution (Fukuoka), 99
open-pollinated seeds, 70
organic agriculture, 10–11
organic certification, 45–47
"Organic Permaculture Market Gardening
  and Economic Performance," 4
oven, 24

Parisian market garden tradition, 2, 66,
  75–76, 82, 84–90, 94
pekarangan, 134

perennial plants, 71–72, 156, 158
Permacultura criolla, 139
permaculture
  abundance and, 3–4
  Bec Hellouin method, 2–4, 114–115,
    118–126
  biointensive methods and, 72
  biological agriculture and, 10–12
  market-scale, 58–59
  microclimates and, 60–62
  microfarms and, 174–175, 197–199
  nature and, 1, 9–10, 58, 170
  no-till and, 151–153
  overview, 51–54
  productivity and, 121–122
  time and, 62
  training in, 207–208, 211
  worldwide, 138–139
Perrine, ci4, ci7, 22–26, 47, 55, 81,
  101–104, 114, 125
Petit, Elsa, 73
Pierre, Abbot, 19
Pincheloup farm, 44
Pita, 27
Plants for a Future, 133
plowing. *See* no-till agriculture
Ponce, I., 90
ponds, ci14, 55–56, 60
Pouëdras, Lucien, 39–40
Pousset, Joseph, 160
prana, 36–37
processed foods, 35–37
productivity, 65, 70, 76–77, 121–122
profits. *See* income
Pupoli, 7–9, 27
Pur Project, 125

qi, 36
quality of life, 209
Quintinie, Jean-Baptiste de la, 85

Rabhi, Pierre, 110, 224
raised beds, ci14, 53–54, 63, 65, 71, 80, 149

ABOUT THE AUTHORS

PERRINE HERVÉ-GRUYER has worked as an international lawyer and head of the legal department of a major company in Asia, and has volunteered with the High Commissioner for Refugees. When she turned thirty, Perrine radically changed lanes and began taking courses in psychotherapy, specifically in relaxation therapy, publishing a book titled *La Relaxation en Famille*. With her husband, Charles, she created Le Ferme du Bec Hellouin. Perrine also serves as a Green Party representative with the Regional Parliament of Haute-Normandie, where she oversees a committee focusing on agriculture.

CHARLES HERVÉ-GRUYER has been passionate about the relationship between humans and nature since childhood. He circumnavigated the globe for twenty-two years while operating a floating school that focused on ecology and indigenous cultures, the subject of many of his books and documentaries. Charles then directed his research to exploring our inner world, studying psychology, relaxation therapy, massage, and yoga instruction. Anxious to explore the most environmentally friendly farming practices, Charles created Le Ferme du Bec Hellouin with his wife, Perrine, in 2003. The couple then began an experiment merging various small-scale, intensive, organic, and natural agricultural techniques that would pique the interest of European agencies planning food-security strategies and would raise awareness about permaculture in France. The farm also operates a permaculture school.

For an English-language version of La Ferme du Bec Hellouin's website, please visit http://www.fermedubec.com/en.

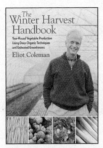

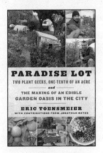
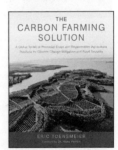

the politics and practice of sustainable living

CHELSEA GREEN PUBLISHING

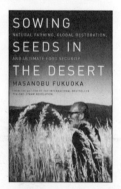
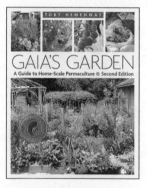